AF572700

THE DESIGN CONNECTION

ENERGY AND TECHNOLOGY IN ARCHITECTURE

THE DESIGN CONNECTION

ENERGY AND TECHNOLOGY IN ARCHITECTURE

Edited by

Ralph W. Crump • Martin J. Harms

Preston Thomas Memorial Series in Architecture

VAN NOSTRAND REINHOLD COMPANY
NEW YORK CINCINNATI ATLANTA DALLAS SAN FRANCISCO
LONDON TORONTO MELBOURNE

Van Nostrand Reinhold Company Regional Offices:
New York Cincinnati Atlanta Dallas San Francisco

Van Nostrand Reinhold Company International Offices:
London Toronto Melbourne

Library of Congress Catalog Card Number: 80-16500
ISBN: 0-442-23125-3

Manufactured in the United States of America

Published by Van Nostrand Reinhold Company
135 West 50th Street, New York, N.Y. 10020

Published simultaneously in Canada by Van Nostrand Reinhold Ltd.

15 14 13 12 11 10 9 8 7 6 5 4 3 2 1

Library of Congress Cataloging in Publication Data
Main entry under title:

The Design connection.

Essays delivered at a symposium at Cornell University, Nov. 1978.
Includes index.
1. Architectural design—Addresses, essays, lectures.
I. Crump, Ralph. II. Harms, Martin J.
NA2750.D398 729 80-16500
ISBN 0-442-23125-3

To

Preston Thomas

Preston Thomas 1952–1974

Preface

The chapters in this book were delivered as lectures in 1979 at Cornell University. They are endowed by Mr. and Mrs. Leonard Thomas of Auburn, N.Y. in memory of their son, Preston, who died in an accident in 1974. Preston was in his fourth year of study in architecture at the Department of Architecture at Cornell. He was actively interested in the problems of energy depletion and I think it is very appropriate that this book, the first of a perpetual series, be dedicated to his memory.

We are living in a time of retrenchment of the world's resources. It is the purpose of this symposium to explore the implications of this retrenchment on architectural design. Today's societies possess the technological ability to diversify their output with new materials, varied mechanical systems and computerized controls. This has provided an unprecedented opportunity to build urban environments that are responsive to variations in climate and natural environment, and thus will use less energy.

Yet today there are few architects experimenting with alternate sources, allowing micro-climate to influence their design or recognizing the impact of the sun. The few who are translating their technical findings into a suitably built form demonstrate how architecture should develop from the relationship between man and nature. We recognize that there is no set of clear-cut answers to the current and future energy crisis. The technological advocates see the problem as a short term one, using short-gap measures to increase our capacity to produce conventional forms of energy. Within our present, sealed buildings, performance is being improved and the equipment efficiency increased. These, however, are primarily engineering problems —attempts to rectify a major architectural error in judgment.

Others are calling for a re-evaluation of needs and accelerated research programs into solar and wind forces and more efficient use of available building systems. Architects can no longer think of a building as a disposable, static object in space, but rather as part of a growing, changing, continuous process.

The six architects who lectured at Cornell for this symposium are concerned with these issues. They were researching and designing relative to energy conservation long before the current crisis. In these lectures, they discuss "the design connection" and its relationship to energy and technology.

Ralph Crump
Professor of Architecture
Cornell University

Foreword

The connection between buildings and energy is an obvious one. The human body uses energy in a constant process of adjustment to the surrounding natural environment. Apart from a few highly specialized climatic regions of the world, this expenditure of energy is so physically and psychologically taxing during long periods of the yearly, seasonal and daily climatic cycles that an intermediate environment is necessary. This must not only relieve the human body of these burdens, itself accepting much of the energy load, but it must also provide space for human activity.

Together these two requirements are basic to all buildings. Since most of us spend longer hours inside buildings than out, certainly more than in our cars, it is obvious at the simplest level that the way in which we use—and frequently waste—energy through the modifying mechanism of the building is significant in relation to the present shortage of energy resources.

Now, insofar as buildings are fairly simple "shelters" of human activities, this is apparently not a difficult problem. Measures related to a building's construction (such as increasing insulation in walls) or to its use (such as adjusting the thermostat) may be taken with beneficial results. To save even more, we can look beyond the *fact* of the building's role as a modifier of the natural environment to the range of *alternative ways* in which the modification might take place. As orientation, sun angle and wind direction are taken intc account, it readily becomes evident that decisions about the building container itself are as important as the environmental control systems it contains in achieving minimum energy usage. Choice in the size of window openings, determination of building shape and the ratio of enclosure area to internal volume all become critical.

Choice immediately implies the need for judgment, and it is this capacity that is presumably the long suit of the trained designer. The designer must also possess specific knowledge and be capable of reasoning logically in solving problems. In this sense, he is no different from the lawyer or the doctor. Society has a right to expect straightforward competence from him as it does from the other two.

Yet, in relation to the pressing energy problem, society's expectations are not being met. Why? A familiar explanation is that getting buildings designed and built is an immmensely complex undertaking in which the designer is but one of the participants. A reason more directly relevant to the present situation is that architects are as typical a cross-section of society as any other group. Many of them believe the "energy crisis" to be merely a temporary condition. Or one could assume simply that some—maybe many—architects are indeed incompetent. While there is possibly some truth in all these statements, none of them strikes at the heart of the matter.

In discussing the connection between buildings and energy, I pointed out the necessity for buildings to provide basic shelter and space for human activity. But buildings must also operate on other levels to be recognized as architecture. Sir Henry Wotton's definition of architecture—"firmness, commodity and delight"—is oversimplified, but it gets straight to the point. Architecture must possess qualities that go beyond mere convenience and appeal to the human senses and the human spirit. To achieve this essential dimension beyond the convenient, the architect must have at his command the control of "space" and of "fabric"—i.e., the material substance of the building that defines space. Architecture results from the designer's conscious synthesis of "concept," the governing single idea behind the design, and "percept," the aggregated impressions of the way the building looks and feels to the observer.

If all this might reasonably be presumed to be obvious, one can only observe two things. First, that an infinitesimally small proportion of all buildings erected in the second half of this century actually possess spatial and other formal qualities to the degree that "delight" is adduced in the observer—and second, that this is because many architects themselves are either unclear about an appropriate relationship between formal intention and convenient deployment of building structure, or entirely deny that architecture must possess a formal dimension.

Presently, many in the profession are busy laying to rest the corpse of modern architecture either on the grounds that it has failed society or that it is undernourished as a medium of expression. In its place is being erected a somewhat shaky structure of ideas meant largely to correct the latter condition: post-modernism. To the extent that these ideas accept, through a reevaluation of architectural history, allusion, metaphor, memory, complexity, contradiction and the rest, the experiments are admirable. Indeed, at one level they imply the husbanding of resources through the preservation and transformation of existing buildings and environments. In this sense, energy is very much a part of the emerging picture of new ideas in architecture. But to the extent that there is confusion over the central question of precisely what is meant by the statement that "architecture is an art," I believe that a majority of architects are not only unwilling but are unable, to grasp the design opportunities that follow from entering energy into the architectural equation.

So, setting discretion and good sense aside, let us address that elusive question every beginning student asks (and, receiving no satisfactory answer, refrains from further questioning until the point in mid or late career as an architect when he possesses the experience and wisdom to begin formulating an answer himself—yet is by then so immersed in worldly matters that it hardly seems to have been worth asking in the first place): Is architecture an art or a science?

Architecture: an Art or a Science?

Geoffrey Scott, in his *Architecture and Humanism*,* written in 1914, stated: "Architecture simply and immediately perceived is a combination revealed through light and shade of spaces, of masses, and of lines. These few elements make the core of architectural experience, an experience which the literary fancy, the historical imagination, the casuistry of conscience and the calculations of science, cannot constitute or determine, though they may encircle and enrich." He admits the role of technology, history and social and moral values as shaping it into a "useful" art, but warns against interpretations in which these factors play a central role—"the romantic fallacy," "the mechanical fallacy," "the ethical fallacy" and "the biological fallacy."

Joseph Hudnut comes down similarly on the side of art in his 1952 essay, "The Three Lamps of Modern Architecture".** "Architecture exists in a realm peculiar to the arts," he explains. "There the values of architecture are unique to art and to the nature of architecture, being independent of philosophical, scientific and social values." He, also, admits the importance of factors that define architecture's uniqueness as a practical art. "Our minds are not so partitioned or our understandings so free from bias that ideas originating in another realm than that of art do not influence, and sometimes determine, our judgment of architectural values, [creating a] mental climate, as Whitehead calls it, that surrounds and screens our vision, directs our thought." But, like Scott, he identifies a series of interpretations of the nature of architecture that, taken in isolation, he regards as inadequate: "the lamp of progress," in which architecture is seen as a direct agent of human betterment; "the lamp of nature," in which, by analogy with the processes of natural growth, architecture is identified as expressing organic order; and "the lamp of democracy," in which architecture is "a propagandist for all that is excellent in the organization and operation of human society."

On the face of it, the position expressed by these writers would seem, if not to exclude energy from the architectural equation, at least to relegate it to a position of minimal importance. If this function of architecture as an expressive art is central, can there be much significance in the architect's ensuring the best possible environmental performance for his design other than the merely practical and mundane one of cost efficiency? Presumably not. But, on the other hand, what then can possibly be said about the work of Frank Lloyd Wright, for instance, whom Reyner Banham called "the great environmentalist," for there is little doubt that in his case control of climate and command of technological means are qualities intrinsic to designs that are simultaneously works of art?

To obtain any kind of answer to such questions, we need to investigate the historical evolution of the terms we have been using—"art," "science," and so forth—for it reveals some surprises. Until the sixteenth century, both art and science were understood by the architect to be about the acquiring of "skill." The only difference was that "art" was understood to mean those skills that required practical knowledge and "science" those that required theoretical knowledge. In the sixteenth century, a far greater divergence of meaning opened up, leading ultimately to the schism between "art" and "science" that is found throughout western thought today. Scientific "experiments" emphasized the distinction between "facts observed under controlled conditions" associated with scientific inquiry, and "experience" as meaning "everyday knowledge of practical affairs." Thus, by the nineteenth century, all fields, including architecture, accepted the distinction between practice and theory.

Parallel with these shifts in meaning was the emergence, in the eighteenth century, of the "artist" as concerned with the world of creativity and the imagination, in contrast to the "artisan" concerned with the everyday world of skilled work. The eighteenth century revival of historic scholarship in architectural matters, allied with the increased specialization of the tasks of building and the arrival of mass production all combined to confine the architecture to a new area of specialization, that of the "artist." In accepting this role, the architect increasingly denied the place that technology might have in the invention of architectural form, preferring to relegate this to the specialist engineer and contractor. Thus had emerged the essential dichotomy underlying all debate about architecture today, between the fine arts (including architectural design), on the one hand, and the useful (or applied) arts, on the other.

In other words, something has occurred historically that has distorted and restricted our conception of what is meant by the statement that "architecture is an art." Scott and Hudnutt are both right, it seems to me, provided that the dichotomy outlined above is declared to be false. But, even given acceptance of this more catholic—and traditional—view of "art" in the architectural context, it remains to be shown how this can possibly alter our view of energy in relation to design in any practical sense, and how it might help us to make more satisfactory buildings. I believe the key to this lies in the nature of the *process* of architectural invention.

Architectural invention

Let us take alternative propositions about this process: the first, *architecture as a formal language;* the second, *architecture as shelter.* The first of these holds that architecture is a discipline possessing an internally consistent vocabulary of ideas that constitute a language. Technology, environment, social context and human values are separate external factors that condition what is said with that language. In this sense, architectural concepts may be said to exist independently of architectural problems. Architectural products, however, are synthesized within a continuous process of transformation throughout the development of a design in which concepts are acted upon and enriched by these "external" constraints. The second proposition contends that solutions to architectural problems exist inherently within the nature of environmental, cultural and social contexts and available technologies and are in large part determined by them. Architecture results from the application of technology to the satisfying of needs; architectural forms are a by-product of this process, possessing no independent validity.

Now, clearly, the proponents of an energy-conscious architecture have a natural tendency to support the latter proposition because, in its emphasis on appropriate technology, it possesses an ethical component essential to the energy-conscious design argument. Yet, while it represents a perfectly adequate belief system about the nature of architectural *products* and their *contexts,* it is quite inadequate as a description of the process of architectural *invention.* It is really the former proposition that more adequately describes the way in which good architectural ideas are generated. It is inclusive as a proposition, accepting the primacy of architecture as art in the broadest sense, while breaking down the barriers between art and technology by dealing instead with concepts and the constraints with which they interact. It takes full account of cultural and human values.

The design methods associated with architecture as a "fine art" do not permit this inclusiveness. While the notion of a systematic, formal vocabulary may well be employed, it truncates, and even eliminates entirely, the essential process of transformation. In this sense, it is "formalistic." In the eyes of its proponents, "energy-consciousness" is too heavily laden with social and ethical imperatives to be worthy of consideration. Yet, if this dimension in design problems is ignored, the resulting architecture can never aspire to the broader definition of art discussed earlier.

Book contributors

To consider the roles of energy and technology in architecture, a symposium was held at Cornell University in November 1978. The essays in this book were all delivered as part of that symposium (which was called "The Design Connection—Energy and Technology in Architecture"). They are admirable as clear explications of strongly held views on energy and design, and they are particularly interesting in the light of the thesis developed in this Foreword. From reading them, it could be inferred that their authors occupy a position more or less opposed to the idea that the central concern of energy-conscious architects should be that of design vocabulary and language. But in fact, though they might well disagree with the emphasis, I do not think they stand in opposition. The source of the apparent difference is that their contributions are almost all about matters of human and cultural values that can provide meaning to architectural forms rather than about the nature of design in the abstract.

As Joseph Hudnut says of his three lamps of architecture, what they have to say "are one and all regenerating influences—the suns before which pretense and cant have fled, new truths have been revealed and a new progress has been made possible." Indeed, representatives of Hudnutt's categories can be identified from amongst the symposium's contributors.

Richard Stein, in his ringing attack on revisionist cant, his appeal for a return to rational thought and his recognition of a "design imperative" imposed upon us by the needs of the future, represents the "lamp of progress." Sim van der Ryn sees architecture as an analogue of evolutionary processes found in the natural world and clearly stands for the "lamp of nature." Ralph Erskine's passionate insistence on the social mission of architecture leaves us in no doubt about his role as the "lamp of democracy." However, these three are no mere commentators—they are also designers. Even if they omit such references in the written word, they demonstrate in the quality of their architecture (some of which is illustrated) not only a mastery of environmental and technological factors, but also the importance of formal clarity and consistency.

Cesar Pelli occupies a special position in the group. He is one of the lions of the profession at the present time. His sleekly engineered buildings attest to his fascination with the iconic possibilities of modern technology, possibilities he has exploited with flair and imagination. Up till now, however, he has not been numbered amongst those whom one would immediately associate with energy-conscious design principles. Yet his contribution to this book indicates considerable interest in these issues, and a concern to give them even greater expression in this future design work.

Ralph Knowles also occupies a special position, but for a quite different reason. His investigations into the relationship between energy and form commenced well over a decade before there was any hint of an energy crisis. His work has been carried out without fanfare and with the frequent involvement of students. It is tempting to classify it as research, but really it eludes classification. It deals with matters of locational, climatic, cultural and historic content. It involves quantitative criteria, yet is concerned throughout with issues of "quality." Its apparently mundane present preoccupation with constructing a sun rights code for cities is deceptive. He is actually about the business of stating a new architectural paradigm that could have the potential for transforming our physical surroundings according to the deepest human values rather than those of the market place. He convincingly makes "the design connection."

James M. Fitch opened the symposium. As an architectural historian, he has been the keeper of the environmental conscience of American architecture through three decades. His re-visit to "American Building—The Environmental Forces that Shape It" is entirely appropriate. It coincides nicely with the present resurgence of interest in architectural history, but warns designers to be wary of treating history only as a repository of architectural forms. It also sets the stage for the other contributions by stating in straightforward terms what it is absolutely necessary to understand first about the relationship between man, nature and the environment of buildings before we can go on as designers to make environmentally responsive architecture.

The six contributors have quite diverse things to say, but they are united by one fact. If the energy crisis disappeared tomorrow, they would not abandon their point of view. In each case, the point of view is one unconcerned with expediency, gadgetry or band-aid solutions. It is concerned with design in the broadest and most fundamental sense, seeking constantly to "enrich" architecture in ways to which people can respond and so enhance their own lives.

Martin J. Harms
Director, Architecture Program
School of Design
North Carolina State University

Contents

THE DESIGN CONNECTION

ENERGY AND TECHNOLOGY IN ARCHITECTURE

1. Introduction

James Marston Fitch

Although the participants in this seminar on architecture and energy are very different people, psychologically and culturally, we have shared for a number of years a concern about the direction of architecture.

Decade after decade we have been part of a radical minority which has challenged the conventional wisdom of the field. If, indeed, history is catching up with us, it will be a cause for satisfaction on our part. We can only hope it overtakes us while we are still in this earthly vale of tears, so we can be present to see recognition of some of these concerns—especially by the architects of the western world, and most critically by the architects of the United States.

It seems to me—and I think I speak for my colleagues—that a special set of historical circumstances has enabled American architects to operate on an "Alice in Wonderland" set of assumptions; namely, the assumptions that energy is cheap and that energy is inexhaustible.

With these two kinds of fundamental misconceptions of experiential reality, it has been possible for the architects of the western metropolis, and especially American architects, to behave in a fashion that will soon be seen (if it is not already) as authentically hallucinatory. It is obvious, of course, that energy was never cheap. It begins to be obvious that it was never inexhaustible. It has only seemed cheap to Americans because of a very peculiar set of circumstances. We fell heir to a fantastically rich, primeval patrimony possessing all kinds of resources: wood, coal, steel, water power, iron, bauxite—all the raw materials of architecture. Because of this legacy, it is easy to understand how the misconception that our energy resources were inexhaustible could have been propagated most ferociously in America.

But the scale of our waste of energy—our profligate use of energy—actually has the curve of a logarithmic, not arithmetic scale, especially since World War II. The illusion that energy in all the forms that go into architecture was cheap was based on several factors. In addition to continental resources, we have had access to those of the rest of the world. In short, the true cost of energy was being absorbed by what we now call the Third World. If aluminum was cheap enough to replace glass, it was because it was produced by the underpaid people in the Caribbean. Or, if heating oil could be sold for 11¢ a gallon as recently as five or six years ago, it was because the real cost was being absorbed by underpaid Venezualans and Arabians. If we analyze raw materials from this point of view, we will see that the cost was never cheap. It was only cheap to Americans, because they occupied this unique position in Wonderland.

Another reason energy has seemed cheap to us is that we have not yet had to face the consequences of our obscene misuse of it, the true costs of which are going to be absorbed by our children and grandchildren. It is they who will handle the question of so-called "disposable" cartons and packaging, most of which, of course, are not biodegradable or chemically disposable at all.

The larger, more terrifying prospect of what we are handing on to the future is atomic energy—the question of how wastes from this process are going to be handled. If atomic energy can be described as cheaper than hydroelectric energy today, it is exclusively because the true societal costs of atomic energy are not being faced today.

One could go on with this catalogue of horrors indefinitely. What is becoming clear to all of us, even architects, is that this Alice in Wonderland period is ending. I personally think that it has already ended. This is going to force upon all aspects of American society—and especially on the construction industry—a radical revision of attitudes toward reality, toward the built world which we inhabit. It is going to compel us to realize that this built world which we handle with such profligacy (those great cities such as Chicago, Cleveland, New York and every other big city in America) has enormous areas of urban tissues which we have discarded in much the same way as we discard milk bottles or Kleenex. We've built these great cities; we've lived in them briefly; and then we have proceeded to discard them, moving further and further out into the countryside. Among other forms of waste, we have despoiled critically valuable agricultural land in this process.

If my understanding of the situation is correct, we are going to be compelled to re-evaluate these used, so-called obsolete artifacts of our culture, whether a milk bottle or a St. Louis suburb. We are going to have to realize that our present attitudes represent a grotesquely inaccurate understanding of the long-range societal value of these artifacts.

The whole concept of technological obsolescence is an invention of the last 50 years. Fundamentally, it is an American invention, based on the miscalculations of the availability of energy. There are, of course, certain areas of American technology in which there have, indeed, been very rapid and accelerating progress. This would be true in science and the higher technologies, particularly in space exploration. The Sputnick of 1956 might very well be authentically obsolete by 1968. But this concept, which applies only to very special and discrete areas of activity, has been extended to cover the whole range of human artifacts. Detroit is the most bizarre example of this incorrect application, with its constant insistence that we purchase the new models of automobiles. For half a century, Detroit has been hard at work trying to convince us that change is identical with progress, and that anybody who dared to challenge this concept is automatically reactionary. Year after year, "new models" are rolled out for our purchase, although it is obvious that they do not represent any qualitative advances over last year's models at all. In this case, technological obsolescence is a farce. A Chevrolet of ten years ago is just as good—possibly even better—than the latest model.

This criterion of technological obsolescence has been accepted also by architects and planners—indeed, by the whole construction industry. We've uncritically endorsed the proposition that anything that is new is automatically better than what preceded it. This has led to the situation I speak of: millions of acres of expensively developed urban tissue which have been briefly used and then discarded in favor of new and reputedly better construction.

Historically, this is an unprecedented attitude. Until the last century (possibly until just the last 50 years), people recognized the enormous amount of energy that was required to build anything, whether it was a quilt, a pair of shoes, a house or a town. It was commonly understood that these artifacts were deposits of what Richard Stein has called "embodied energy." As a consequence, they were assets, those old shoes and houses. They were not obstacles to progress, but part of the fundamental resources of the society. This meant that no artifact was thrown away until the last possible ounce of energy had been extracted from them. A typical American artifact such as a quilt is a quintessential demonstration of this process. Any American with his wits about him in 1800 knew how much total energy it took to produce a yard of cloth: to grow the cotton, beat the hemp, raise the indigo, spin the twine, weave the fabric and cut and sew it. Everybody understood that implicit in that yard of cloth was a great deal of energy, both human and mechanical. As a consequence, it was handled with great care, even by the prosperous. A pair of pants worn by Johnny was then passed on to Johnny's little brothers until the pants started to disintegrate. Then grandmother cut the pants into pieces and made a quilt of them, extending even further the life of the fabric. Finally, the fabric was either made into felt or converted into first class paper, and thus recycled again.

This acknowledgment of the costs of energy was true not only for the peasant, but for the richest and most powerful sections of society. The Roman emperors, for all their fantastic access to energy, actually built comparatively few brand new constructions, and they threw very little of what they conquered away. They did build new highways and new ports when it was necessary for the expansion of the empire, but fundamentally, the Roman emperors were compelled to recycle their resources in much the same way as peasants did. If you look closely at the Roman Palatine or the Forum, you will see that even in such Imperial artifacts there was a continuous reworking of used material. If you examine the base of the Acropolis carefully, you will see that most of the retaining walls are constructed with the drums of discarded columns. Whole streets at the base of the Parthenon are likewise paved with discarded columns.

Historically, it is perfectly obvious that it has only been within the past 50 years—and even then only in North America and western Europe—that a wasteful attitude toward energy has been possible. It wasn't that our ancestors were smarter than we are; it was merely that there were severe parameters, insuperable limitations, to what was technically possible for them. This is what kept our ancestors "in shape" and made their attitude toward the built world more rational and sane than ours.

I, for one, am personally encouraged by the "energy crunch." Far from being distressed by the coming shortage of energy, I think it's going to prove to be a very healthy, therapeutic development. It will force us to examine the real world in a way that we haven't done for decades.

The members of this symposium have long been concerned with establishing sensible economic parameters for the use of energy. From Knowles' long exploration of the architectural consequences of sunlight, the solar energy that falls on all buildings, to Erskine's theoretical investigations of technology and the Arctic way of life, all of us have in one way or another been dealing with the architectural and urbanistic utilization of energy. Knowles has pioneered most recently in his work in solar access zoning—i.e., establishing design criterion whereby the whole fabric of a city can be zoned in such a way that every square meter of that city would get an equal share of the sun that falls upon it. Erskine's work demonstrates the special ways in which energy can be applied to make life not merely possible in the far north, but also more amenable. As Erskine has pointed out, the north of Sweden and Norway have been populated by the Laps quite successfully for millenia. Lapland culture was of a much more primitive level than that required by modern industrialized society, but Erskine has demonstrated the fundamental sagacity of Lapland procedures. Though it is not possible for modern Swedes to live in igloos, he argues that it is also not possible for modern Swedes to ignore the precision and brilliant lucidity of the igloo as a means of exploiting the natural environment.

I have thought of myself—fatuously, it now appears—as being an avant-garde theoretician of energy conservation. At a recent conference organized by the distinguished physiologist and environmentalist, Rene DuBois, I learned that whole circles of specialists are far ahead of me. An ecologist from Rutgers University, for example, advanced the proposition that it is no longer enough for the human race to decide if energy is clean or not—whether it is generated by "clean" waterfalls or "dirty" coal or atomic energy—but that it is necessary to see that above a certain point *all energy threatens ecology.* All ecological systems have a delicate equilibrium: the application of energy beyond a certain level, no matter what the source of that energy, would be not only undesirable but would be dangerous to survival of the system itself. From evidence such as this, which is now coming in from all levels of society, it is clear that a decision is going to be forced on us, whether we like it or not.

If we architects try to be more precise about this problem, we must realize that we are talking about two forms of energy: animal (especially human) energy and mechanical energy. Both forms are involved at three levels—construction, maintenance and habitation of built space. The first is the energy required to fabricate the building—not only the energy required for construction in the conventional sense, but *all* the energy required, right back to the extraction of the raw materials. It is now possible to quantify this total for the first time, to understand in terms of BTU's how much energy is embodied in any building. The second, more familiar, level is the use of energy in the maintenance of buildings; that is, once the buildings are constructed, how much energy will be required to keep them in operating order? My own special interest has been concentrated in the third level—on the amount of energy it takes to occupy space and to perform effective work in that space, and on how to analyze the role of the building in which the work takes place (how it conserves our own physical and psychic energy in performing that work).

In his primal state, man is catapulted into a raw environment. There, a single interface exists between him and nature—his skin. In most parts of the world, this places a great load upon him. In fact, there are very few climates in which man could survive without the mediation of clothing or buildings. Hawaii would be one of them, but even in that genial environment, libraries, schools, hospitals and all the processes of civilization require buildings. As a consequence of this necessity, man has invented architecture. Whatever higher levels of satisfaction architecture may afford us, its fundamental task is to create a new environment, a middle scale between the microenvironment of our bodies and the macroenvironment of the world, to manipulate the flow of energy in our favor. If we begin to apply criteria like this to the critical analysis of architecture, we begin to see how often modern technology permits us to err. Many very sophisticated contemporary architectural constructions perform with grossly less efficiency than the Eskimo's igloo. One could say with complete confidence that, compared to the igloo, the great glass tower of the Hancock building in Boston is a primitive construction, if one takes as the criterion of architecture the wise and sane application of energy for human purposes.

It is in such a context that we will have to discuss energy in the future. Thanks to the work of men like Stein, we have the means of re-evaluating the built world. Before we decide to construct a single new building, we must re-evaluate all existing building stock around us. Before we destroy old buildings in favor of new structures we must understand that this not only involves the expenditure of a great deal of energy, but might very well be less efficient. It is in this context that the historic preservation movement seems to me the most encouraging development in the American scene today. The average American citizen, without the assistance of professionals, has come to the conclusion that the built world can, and must, be conserved. Here at Cornell University, we can see how the slide of downtown Ithaca toward oblivion has been altered and reversed in the last decade, largely by citizen activity. This activity can be seen all over the country. And, the citizens are ahead of architects, planners and engineers. A citizen's motivations may not always be "scientific"; he may want to save an old street because his grandmother lived there, or an old house because George Washington slept there. Motives are often mixed. But, by and large, the public is ahead of the "experts" in understanding the problems of conservation of energy.

Young architects may tend to see my description of the future as a dismal one, as though I am suggesting that they have no future except to knit and patch obsolete fabrics. This is wrong in two respects. First, a surgeon would not consider it a waste of time to knit and patch damaged parts of a patient's body! A surgeon would consider it a noble and creative function. Why should the analogous level of activity be demeaning for architects? The whole history of architecture, including the noblest architecture, indicates that the reverse is the case. And second, men such as Ralph Knowles try to convince young architects of the critical necessity of paying attention to the solar energy that, whether they like it or not, will fall upon their buildings. They tend to think that Knowles is trying to box them in, that he is giving them a limited range of mechanistic parameters which will restrict or cripple their creative ability. This is a profound misunderstanding of the thrust of Knowles' work. In analyzing even one problem—such as how to manipulate the flow of solar energy between outside and inside, Knowles offers a palette of aesthetic choices incomparably wider than current formalistic architects enjoy today.

If young architects with artistic ambition really responded with wisdom to the force of the sun, they would see an infinite range of possibilities opening up before them. They would understand that the architect doesn't have to *invent* avant-garde geometries: the sun itself would dictate those shapes. Rather than imposing a brutal limitation on creative potential, I think I am suggesting the only real avenue for it. Stravinsky, who was an authentic avant-garde in his field, an innovator who changed the whole course of modern music, said at the very beginning of his artistic career, "When all things are possible, nothing is possible." I think that is true for architects and the built world. It is not our task to invent novel forms that merely titillate our egos for the moment. It is our task to respond with widsom and discretion, using all the resources of modern science and technology, to the real experiential needs of the people. If we do this, architecture will be not impoverished, but vastly enriched.

2. Energy and Technology - the Design Imperative

Richard G. Stein

We are in a period of architectural evaluation brought on by dissatisfaction with a great deal of the post-World War II architecture and its effect on cities, suburbs, the buildings' users and our patterns of energy and resource usage. The discovery that the building industry (and I include the architects as the design component of the building industry) has not been responsive either to human or to environmental needs has led to a critical free-for-all in the architectural press as well as in the profession at large.

All of us are products of our times. I received my educational training during the 1930's and studied and worked with Gropius and Breuer for several years in the late 1930's and early 1940's. As a result of that experience, and possibly the time in which I was embarking on my architectural career, I was convinced that the new architecture was the expression of a growing clarity and reasonableness in the way people saw themselves, and that this basic rationality, coupled with the study of how buildings perform, was the generating force from which architectural form derived. While the architects of my generation had a deep respect for the architectural work of such men as LeCorbusier, Aalto, Scharoun, Lubetkin and a number of other architects, and even though a great deal of our own work was really suggestive of the forms developed by some of these men, we did not consider that we were working in their style, but rather with an open-ended philosophical approach. We were also convinced that because of the factors that determined architectural decisions, the new architecture had a strong social content that would eventually make it the obvious choice for schools, houses, residences and all the other buildings where people's daily activities took place.

At that time, and continuing through the 1940's and into the early 1950's, there were a number of very interesting and useful studies being carried on, and there were also some buildings being built that were certainly adventurous, forward-looking and fundamentally inventive. I am thinking particularly of such buildings as the Ministry of Information Building in Brazil, by LeCorbusier, Costa and Niemeyer; some investigations in vernacular styles of building carried on by Jimmy Fitch as part of the *House Beautiful* series; and a number of schools that operated very easily with the natural environment, including some important wind study tests by William Caudill of naturally ventilated schools.

By the early 1950's, there was already a new concern sweeping through the architectural com-

Public School 55, Staten Island, New York.

The west facade of the school indicates the approach to lighting and ventilation. Projecting windows at the front of classrooms admit light which sweeps across the chalkboards. The horizontal strip of windows admits light and permits views of the outside for the students seated in the classrooms. The lower floor contains kindergartens. Since these have an informal seating pattern with no front, back or sides to the classroom, no chalkboard illumination is required.

A detail of the facade from the north shows the opaque side of the projecting triangular prisms. These eliminate the glare that is often found where the chalkboard meets the window.

The north facade uses windows in a similar manner. The structural bay in the upper floor is the library and needs no projecting window since it has no chalkboard. The lower floor contains offices. The window type provides maximum opening for introduction of outside air through projecting windows at the top and hopper windows below for draft-free ventilation.

munity, and the offices which were to grow into the largest architectural offices in the country. These offices came to the conclusion that if growth was their objective, the way to achieve it was to drastically simplify the design procedure and the building procedure. There were technological options and developments that could be exploited to permit this change in office size and responsibility to take place. Probably the greatest factor on which the change depended was the development of mechanical components that permitted comfort conditions to be generated almost entirely internally. This shifted the responsibility for actual environmental conditions from the architect to the engineer, and while it presumably freed the architect to pursue new design options, it also stripped him or her of one of the important historic reasons for the existence of architecture as a profession. Once this path was embarked on, it led rapidly into an almost complete dependence on the sealed building; the undifferentiated facade; and mechanical production of light, conditioned air, internal temperature and humidity content. It resulted also in a constant escalation of the amount of fuel required to keep our buildings operating. An examination of post-World War II office buildings conducted by Dr. Charles Lawrence when he was New York City's public utility specialist, revealed that the average energy used by office buildings went up every 5 years, and in the 20 years covered by his study (from 1950 to 1970), the average use doubled. It must be recalled that all of these buildings respond to the same general programmatic requirements for offices, stenographic areas, reception areas, conference rooms and so forth.

While we think of the sealed building as describing primarily the commercial high-rise office building in downtown urban areas, the fact is that it is also the characteristic method now used for building schools, hospitals, many private residences, city apartment buildings, museums, motels and hotels, as well as factories and industrial buildings. The design problem has shifted from one of making the building environmentally compatible, to one of making the skin area as small and as tight as possible to prevent any interchange of inside and outside conditions. Of course, at the same time as the face of the building is considered impervious, there is somewhere in the building an enormous air intake that brings in all of the air that then has to be modified for the temperature and humidity requirements needed in the building. And since this air has to be distributed through a system of ducts of restricted dimension, the temperature has to be either raised or lowered substantially beyond what is actually required, since it will later be diluted as it mixes with the air in the space. We will come back to the implications of problems like this a little later. However, it should be understood that I have just described the principal design approach used in many offices and either taught or tolerated in most schools.

Through this period, particularly in America, the design procedures described above were the agents that produced a building's design. I would like to recall an incident that took place in my own office in the mid-1950's. At the time, we were working on a large hospital. While laying out the floors, I was concerned that the patients and the people working in the building should not be overwhelmed

Vacation House, Martha's Vineyard, Massachusetts.

The southeast facade of this vacation house indicates the variety of openings and skin treatment used to introduce light, air and solar heat and to prevent the entrance of insects. The center section is screened on the two open ends and overhead. The bedroom section, on the left, has screened slot projections that can be closed from the inside with hinged plywood doors, and, on the other facade, sliding wood panels that cover screened areas. Light is introduced through fixed glass. In the bathroom and small bedroom, light is introduced through light scoops in the roof. The right-hand wing, containing a living, dining and kitchen space, introduces light through fixed glass and air through large screened openings. These can be closed with sliding barn doors.

The northeast facade opens to the outside through screened doors. The openings can be closed off through the use of sliding barn doors. The entire construction has low mass, just the single skin on the outside of the studs. Insulation is provided above the plywood roof sheathing, below the roofing.

and depersonalized by the endless repetition of identical spaces. I therefore instructed the man who was working on one of the drawings that I wanted the corridor to have the interest and variety—and some of the characteristics of—a street, with certain parts of the corridor opening to the outside wall, with changes in the width of the corridor through alcoves that would serve as pauses along the way, through the use of color and so forth. The man who had just come over to our office after several years at one of the largest and most respected American firms was puzzled and a little shocked by my sketch of what was to be done. He informed me that at his previous office, that would never be tolerated, that once standards were set for a corridor, a window component or a partition detail, these were not to be changed under any circumstances, and they were then repeated ad infinitum over, and within, the entire building. This became both a matter of office efficiency as well as a question of design philosophy.

The absence of particularization, the endless multiplication of the unit, became the sought-after characteristic that underpinned the selection for architectural publications through this whole period. Obviously, it was a kind of philosophy that would permit an office to train an office staff and permit that office to seek and carry out a great number of commissions without having the work-challenging problem of making innumerable design decisions within, and outside of, the building. It permitted the refinement of curtain wall joints and assembly techniques that could then be applied equally effectively regardless of climate or orientation. As a result, one finds buildings that are indistinguishable in Houston and Montreal, in Minnesota and Hong Kong. As this occurred, the major differentiation between buildings took place in the size and requirements of the mechanical plant. And even here, mechanical plants were not fundamentally different in different areas, but only larger or smaller. No effort was made to develop minimal mechanical plants directly responsive to the different load conditions put on them by different kinds of buildings in different parts of the world. Most plants are still substantially overdesigned, with the assumption that if one has to combat varying natural conditions, the way to do it is by having enormous reserves of power, in most cases the power to provide cool air, with an additional energy reserve to cancel out some of that power if it is not needed, through the introduction of heating coils or elements to raise the temperature of the chilled air.

So, in a period of about two decades, the entire approach toward building design had substantially changed. At the end of the period, the product had very little resemblance to the product at the beginning. Since the public rationale for the change could not be expressed purely as a business administration decision—that it is primarily a more convenient design idiom for turning out large quantities of work—there was a search for acceptable precedents and the development of a language that would be more acceptable to those concerned with the human spirit. Probably Mies van der Rohe's skyscraper study for a glass building in 1921 came closer to furnishing this rationale than any other single precedent. While we think of Mies as the very sober, reasoning exponent of the fully and uncompromisingly understood technical building—made possible by the freedom created by industrial means, and raised to a level of perfection by exquisite detailing and proportion—in truth, his concept was romantic in that basically the building was more a *metaphor* of the new era than a *product.* The underlying promise of the glass building was that it was a minimum diaphanous intrusion into nature, that it was so light and insubstantial that people could be almost floating above the ground with no apparent separation between themselves and the outside. The planes of the floors and the light vertical columns would be the only visible intrusions in this containment of outdoor space, and were it not for the presence of the glass, it would be hard to say where the indoors stopped and the outdoors began. The impact of the sun, the stratifying of heat, things that were known and controlled rather carefully by Paxton in the Crystal Palace and others who came out of the greenhouse tradition of early nineteenth century England, were not part of the considerations that proposed this as the ultimate building form. Some 50 years after the original statement was made, the all-glass building is a reality, with many demonstration projects all over the world. It has turned out to be far different from the poetic promise of the early Mies projects. If anything, the amount of sheer brute intervention required to maintain satisfactory conditions in these buildings is greater than in any other buildings that have been built. Throughout the day, lighting is provided right up to that glass curtain. People on the west side are broiled in the afternoon sun, while people on the north side may be chilled. If there is an effort to reject the sun (which flies in the face of the orginal premise), through the use of reflective glass, the result is not a transparent prism, but a great solid mass whose exterior is clearly defined but which substantially portrays everything around it as a billboard might.

It was the dissatisfaction with the performance of post-war buildings, and the boredom resulting from the acres of rectilinear curtainwall grids, that spurred a rapid desertion from the ranks of those who had most recently been identified as the avant-garde of the modern movement. Since the purported generating point of view had been a functional one, the reaction was more often than not an anti-functional one. It is possible to identify some of the major categories of the new dissident sects and see where their philosophies are leading them.

Probably the first group to consider is the one that has discovered the variety in building done without the participation of the architectural profession. The variety, the disorder and (eventually) the irrelevance of the building become the most important characteristics to imitate. However, since the architects who have discovered this new world of random forms, shapes, colors and openings are highly sophisticated critics and practitioners, it is not possible for them to achieve the results through naiveté and innocence. And yet, the forms cannot be copied directly. As a consequence, the forms are used as references and the randomness is exaggerated, but in order to make it clear that this is something different, the forms are now changed in scale or put into illogical relations with one another. The original is used as a historic reference, and the copy has all the innocence of Marie An-

The Wiltwyck School for Boys, Yorktown Heights, New York.

The gymnasium building is one of 13 new buildings that make up the campus. This treatment center for emotionally disturbed children uses a system of prefabricated wall cladding and windows that is designed for flexible response to the differing requirements of the spaces and orientations of the various buildings. In this case, light and air are admitted from above the sight level of people using the gym. The basic source of light is from the two lateral walls of the gymnasium. All windows are modular and combine opening and fixed elements within preset dimensional constraints.

The interior view of the gymnasium shows the quality of light and the relationship of windows and insulated pre-cast panels to the basic structural system used for all the buildings.

In the background of this photograph is the west wall of the gymnasium building. Heat-reducing glass is used on this facade. In front of the lower section of the gymnasium, the locker rooms are lighted and ventilated from high windows. The foreground building contains classrooms. The projecting boxes are small greenhouses on the east and west facades of the building, one to each classroom.

The clinical program building houses a number of individual offices that can be flexibly changed as programmatic requirements change. In addition, since the school is also a teaching and training institution, there are group therapy rooms in the center of the building surrounded by observation rooms. The interior rooms are lighted from above through the triangular light scoops, visible over the left hand portion.

A bay window and window seat in the academic school building accommodates small groups in special teaching situations. It also introduces a differentiation into the building that responds to programmatic requirements.

toinette's peasant life at the Petite Trianon at Versailles.

It must be noted that the elements that have been selected as the basis of this new digression are not in themselves either irrelevant or unimportant. Rather, the shortcoming is in isolating the identifying characteristics as forms and using these as the formal precedents for a new style, a style identified by motifs rather than by philosophy. Not too far away in basic attitude is the allusive historicism that has recently surfaced. If we wish to see an example by a most accomplished architect, we must turn to the new American Telephone and Telegraph building of Philip Johnson's. Here, the formal reference to a Renaissance pediment is included to provoke a favorable response by association. By not exactly imitating the manner and vocabulary of the late Beaux Arts buildings, the architect can remain aloof as an outsider and a critic, rather than as an involved or committed participant. It eases the assessment of responsibility and allows a rapid and polite disengagement if it becomes advantageous to do so, since the gesture can be dismissed as a passing witticism. In both cases, the Las Vegas school and the Chippendale school, the impact on architecture generally is to identify the architect as a non-participant in the production of buildings designed to answer human needs. These answers to real problems are left to others. Problem-solving has become a pejorative term. The building's main purpose has become a didactic one, replacing the printed word or the book of photographs as a comment on the problems and foibles of our society.

Another avenue of response to the blandness of a large part of our most widely published work of these last couple of decades has been the investigation of form as a manifestation disconnected from any of the factors historically associated with the development of form—plan requirements, materials, orientation, wall function specifications and such. The building becomes the medium for the investigations of formal problems, but ultimately, the issues addressed are non-architectural, such as the complexities of double layering of surfaces or the ambiguities of meaning embodied in walls and partitions. The destruction or distortion of information concerning the building itself—what is structural, what shafts enclose columns or pipes (or nothing)—has been an inevitable consequence. The language used to describe buildings is the language of art criticism and the building serves primarily as a vehicle for examining the problems of the artist. Moreover, buildings are often invented in order to pose problems of spacial or plastic derivation. Ultimately, the architect becomes disconnected from the performance of the building. This attitude leads very quickly to utopian projects, to projects which are never intended to be built and to projects for which the graphic representation serves as adequately for these investigations as the building itself. There have been many provocative and interesting models and drawings made and volumes of words describing the issues investigated. If the ideas can be addressed effectively by examining drawings, why bother with constructing the building? In the final analysis, not even the ideas themselves become the end of the

Intermediate School 183, Bronx, New York.

Within a strict structural modularity, the facade of this building introduces openings as required by the organization of the school itself. The perforated reinforced concrete bearing wall (there are no wall columns) is internally insulated. This permits the retention of internally generated heat, whether from the heating plant, the occupants or the lights. At the same time, the thermal mass of the building delays the impact of exterior spring, fall and summer heat loads. Where there are nonrepetitive units, as on the north facade of the building, the varying requirements are expressed on the facade.

Intermediate School 183, Bronx, New York.

Within the courts of the building are sculptures in concrete by Constantino Nivola.

studies, but the drawings of the ideas. Now we are beginning to find that architectural drawings are the end product of the whole process. The drawings become art objects and can then be examined and discussed entirely within the critical framework and with the terminology of art criticism. Many are exquisitely beautiful drawings and certainly contribute to our aesthetic resources. As a major current for the production of architecture, however, I believe the whole movement is not only deficient in its consideration of major architectural problems, but becomes an easy and beguiling alternative to the truly complex examination of all architectural problems.

It is interesting that, for some members of this architectural community, the works of Le Corbusier have been the jumping-off points. LeCorbusier's formal invention is as remarkable as his clear understanding of the limitations and opportunities contained in building with modern technological means, with major social programs and with specific topographic and urban references. As in the case of the groups that have learned from Las Vegas, the form that has served as a model for their investigations has been separated from the conditions that created the form, and in that separation, the most important aspects of the form have been lost.

This fleeing from accountability and responsibility may be characteristic of the times in which we live. It may be that the only survivors will be the huge corporations, so anonymous that responsibility is hard to pin down—and, if it is pinned down, an individual or group of individuals can be sacrificed without in any way affecting the direction or corporate stability of the larger enterprise. The recent revelations of extensive corporate bribing and corruption of officials at home and abroad, and the few cases of retribution, are examples. In some cases, minor executives have been identified and have paid fines, lost jobs or been transferred, but there have been no more fundamental changes than that. On the other hand, individuals can find that a small miscalculation in a complex job can cost them enormous sums in liability claims and possibly the loss of their practices and the loss of any economic stability they had achieved for themselves. Among our fellow professionals—the doctors—there are many who find that with the high cost of their insurance, not unlike the rising cost to the architectural profession, they can no longer afford to practice medicine and therefore decide to devote their full time to other activities. In this atmosphere, the withdrawal from areas that involve responsibility and potential liability is, understandably, sought. It may be that part of the presentation of straightforward information about buildings is becoming more and more fraught with personal danger, and this makes mysticism, withdrawal and non-building especially attractive.

A variation on the theme, I would say, is the appeal of the mirrored building that I mentioned earlier. Not only does it attempt to conceal all information about the building, but its further objective seems to be to confuse even the information that one ordinarily expects to receive from a curtainwall related to its construction system. In some of the newer buildings, and I am thinking of the UN Hotel in the East 40's in New York City, even the divisions of the curtainwall no longer refer to the structural grid behind it. Sloped planes of reflective glass are introduced in a further effort to destroy any references to the true nature of what takes place under the skin.

At the same time as all of this critical attention has been centered on the non-building aspects of architectural practice, there has been a small but growing counter-force that sees architecture as serving the historic role encompassed by the terms, commodity, firmness and delight. This brings me back to my starting point. For the past 8 or 10 years, I have been identified with the examination of energy use in buildings. I had gotten into this involvement primarily because I was disturbed by the gap between the ostensible basis for buildings—that is, that they were designed to respond to various conditions and to perform predictably and well—and the actuality of their failure in most of the field, including satisfaction of user requirements and efficient performance as man-made objects. I put together a number of statistics from 1971 as my part of a symposium on the implications of energy use, conducted by the American Association for the Advancement of Science at their annual meeting. As a very large contributor to national energy use, some 33 percent in their operation alone, the whole assemblage of buildings merited detailed attention. In retrospect, and possibly as something to look forward to in the future, we can relate a number of our unresolved national problems to the deteriorating world trade situation faced by the United States. The amount of American money going abroad (in other words, the inverse balance of trade) has a direct connection with the deterioration of the American dollar worldwide; and from that point of view, the 33 percent of our total energy use required to operate our buildings does have to assume its causative role. Thus, energy use as a measure of building design removed the question of building appearance from the seminar rooms to the halls of Congress.

When we know that buildings can operate equally well with substantial reductions in fuel requirements, when we know that new buildings can be built to perform in a satisfactory manner with even less energy than the reduced levels needed for existing buildings, when we know that the architectural act of designing buildings has a direct connection with the performance of the buildings, and when we also know that there is an inseparable bond between the elegance and tautness of a building and the way it performs, it becomes clear that there are major new fields for architects to move into. It is my belief that we were beginning to approach these fields 25 years ago, but that we were turned away and, instead, concentrated on a kind of architectural approach which has turned out to be destructive to our cities, to our national economy and to the role of the architect.

I believe it is timely to concentrate on the resolution of the real problems that must be faced and not to divert our own attentions into fields of inquiry that will remove us further from these important considerations. What this means is the re-study of various ideas and principles that we had considered either unimportant or solved. Some of these are now being looked into at a small study group that I have set up at Cooper Union. I have been involved with the design studio course there for many years but felt this year that the most in-

Reiss House, Fire Island, New York.

This simple summer house, on a dune facing the ocean, uses glazed sliding shutters in front of screened openings to change from openness without visible windows to a tight enclosure when the weather requires it. It has no interior wall or ceiling finish. A light-colored roof and roof insulation prevent the horizontal surface from becoming a radiant panel. The low-mass walls do not heat up or retain heat.

The typical wall opening arrangement is a glazed sliding shutter that closes a screened opening. The screen is hinged to allow access to the sliding shutter.

formative and useful study that could be made in the design area was a fundamental investigation and clarification of aspects of architecture that we have considered axiomatic but which in reality are not that well understood by any of us. Among the subjects to be examined closely are the performance requirements of buildings (and this includes not only physical performance, but psychological performance and the effectiveness of space provision and utilization—the various performance characteristics one should expect of skins of buildings in relating conditions inside the building to those occurring on the outside of the skin, which implies methods of preventing heat transfer when this is necessary and of permitting heat transfer either from inside to outside or outside to in when that exchange is desirable; the method of introducing and controlling light from the outside to the inside under different working conditions within the building and with different orientations outside; methods of proper introduction of air from the outside to the inside and vice versa, through the skin of the building, propelled by the enormous energy of wind movement and temperature differentials; and provisions for the interception of such unwanted aspects of the outside environment as particulates, noise, excessive moisture and pathogens); the compatibility of provisions for controlling performance factors and the provision of interchangeable components if the performance requirements change; and what is truly meant by grids or organizing systems for different kinds of building, going back once more to reacquaint ourselves with the size of a building system that is appropriate for that particular method of building.

There is nothing proprietary about either the information we need or the search for it. The enormous amount of information required is probably measured only by the enormous commitment toward the use of such information if architecture is to survive and serve its traditional purpose.

3. Integral Design

Sim Van der Ryn

We are ready for a new paradigm. In my 25 years in architecture, I have never seen architecture so intellectually bankrupt. The whole architectural profession seems to have given up its morale and its visionary role in society. And yet we are at a time when what we know and what we can do, as architects, are more important than ever before.

The real task seems to be to translate the rhetoric of the architect as the builder, the integrator and the coordinator, into action. Having spent three years as State Architect in California, I can say that it is possible to do this. It's possible to talk sense to legislators and get them to do some very good things.

The real opportunity we have is the opportunity that results from designing with limited resources, and beginning to become aware of the opportunities that grow out of the depletable nature of our resources. What this means to me is what I am going to define as "integral design," in which architecture becomes not only building, but an orchestration of the other forms of energy that are designed into the form of the built environment. We need to consider what happens to the waste generated by buildings, as well as the energy that is involved in creating a building and the energy that is involved in maintaining it.

The decisions that we make will be paid for by people many years down the road. Our children don't inherit the world from us; we are simply borrowing it from our children. The decisions I make in my work are going to influence those after me, just as the ones you make are going to affect future generations.

We need to look in a more integrative way at how we orchestrate everything that makes up a life support system—not just shelter and the habitable environment, but food, energy, waste and all the things that are part of the system. The kind of paradigm we need to develop is one in which we begin to look more closely at what happens in the natural world. When I went to architecture school—and I suspect it hasn't changed that much in 25 years—I went through 6 years of training and never took a course in natural science. In architectural curricula, you study dead things such as concrete; there is no study of botany or even geology. This missing ingredient was bypassed in the great revolution that took place in architectural education in the 1920's, although the people in the Bauhaus movement were aware of forces that constitute the missing ingredient. They were never integrated into how we think and how we design, and that is the major focus of what I am working on now. So if it seems my ideas relate more to agriculture or biology, it's on purpose. We need to look much more closely at how natural systems work, and we must learn from them.

Integral design is about creating living places or habitats, the humanly created organization of which is analogous to the features of a healthy natural system. By doing this, we can begin to approximate the economy, the efficiency and the simple elegance that is inherent in any natural system. The work we use to describe the tendency to approximate the features of a natural system is "integral"—connected or unified. The dictionary meaning of·integral is "essential to completeness."

It's important to keep in mind that no humanly designed system can ever achieve the organization that natural systems have evolved over millions of years. But we must keep in mind that integral design has as much to do with process as it does with realized form. A house whose shape is analogous to a nautilus shell, or a dome emulating the microscopic structure of an organism, is not inherently organic. *Process* must be analogous as well as shape, although we do tend to think of houses built with natural materials like earth, stone or unsawn wood as more natural than those build with industrial materials such as glass, steel or concrete. Integral design applies the lessons of the biology and ecology of the natural systems to the design of environments for people. This emerging kind of integration of architecture and biology, dubbed "bio-tecture" or "eco-tecture," is in its infancy, although we can already begin to identify principles and patterns.

An obvious question is, "Why emulate natural design?" What is there about the behavior of natural systems that we should pay attention to in designing our cities, towns and houses? The answer is framed by the most natural observed event on earth. The source of all light energy is the sun, but the earth is only habitable through the action of green plants, from lowly algae to towering redwoods, which capture only about one percent of solar energy, and transform it into useful forms of energy for all life. Without these complex natural systems to fix and transform energy, all solar energy would be lost as waste heat and life could not be sustained.

We express this process as *entropy*: the tendency of all energy to degrade into unusable waste heat radiated back into space. The opposite is negentropy, which is the sum total of all life processes which capture and transform energy into usable form. It is negentropy, the work of all the silent plants and bacteria, the entire complex of natural systems, which is the basis for life and civilization, the savings that we accrue through natural systems.

Evolution is a process by which natural systems become increasingly diverse, complex and differentiated, in order to counteract entropy. Evolution, through negentropy, may be seen as nature's slow but certain strategy to achieve stability in the face of the inevitable degradation and eventual death of the planet.

Human beings cannot hope to improve on the efficiency of natural negentropic processes, but we should be able to design habitats and culture in such a way that natural systems, and the information in them, are not degraded. If natural systems are degraded, then human cultural evolution will be degraded and destroyed.

Humans, as a species, are uniquely adapted to

storing information in abstract and symbolic forms, and this gives us our unique ability to manipulate our surroundings as no other species can. It's interesting that some people claim that modern societies are more callous about their effect on natural systems than were earlier peoples. In truth, while there are important exceptions, the rule seems to be that most cultures have been callous about their natural systems and environments. Throughout history, cultures have gained short-term advantage by juggling what ecologists call "early succession-type" monocultures, and as ecosystems deteriorated through early succession-type monocultures, people had to take the consequences and move on. In Edward Hyams' book, *Soil and Civilization*, he documents the ecological destruction of vast areas of the planet by earlier cultures. We don't have to suffer heavy guilt about modern industrial society as destroyer; most cultures have destroyed their environments. So we don't have a corner on ignorance (although we do seem to be in the lead). Hyams points out that most of the cultures he talks about were destroyed by the use of early succession monoculture techniques, characterized by the production of large yields for relatively short periods of time. It's like the story of the goose that laid the golden egg. The keepers, not satisfied with getting the eggs one at a time, killed the goose to get them all at once—only to find that once the goose was dead there were no more golden eggs to be had. That old parable is very relevant today.

What has changed from earlier cultures is our increased potential for the destruction of natural systems provided by a dizzying array of new technology and the increasing insulation of modern urban populations from the effects of ecological deterioration. We have a false sense of security provided by a massive but short-lived supply of fossil fuels—oil and gas, which are simply the stored products of negentropic systems many eons old.

We can do something about this situation by learning how to make the transition to an integrally designed way of life.

It all requires much time to give a definitive picture of what distinguishes a living system from an inanimate one, or how to construct a habitat that fully integrates the inanimate and the living form. At this time, only the vague outline is there, and much of what is learnable must be learned by comparison. We can contrast the properties of the integral system with what I call the linear system, the characteristic early-succession ecology in monocultures that has dominated most human societies. What may look to you like a very complex society, in ecological terms may actually be in a very simple and early-succession stage. In the integral system, energy flows through loops; in linear systems, energy flows in straight lines. In integral systems, parts fit overlapping functions; in the linear system, parts are specialized components. Integral systems have low entropy and low information. In the integral system, memory is stored in many different cells; in the linear system, memory is stored in specialized components—such as architects. In integral systems, there is a high rate of material recovery; in a linear system, there is a high rate of material loss. Integral systems have multiple alternate channels; linear systems have single channels. Integral systems have no waste; linear systems have high waste. These properties exist on a continuum, and we need to develop a feeling for when a particular system approaches the integral or linear.

There are four processes unique to integral systems:

1. *They process materials and energy through closed loops and webs of multiple channels.* The closed loop is absolutely fundamental to the nature of living systems, and many benefits are realized from the intentional application of this concept to the built environment. Living organisms exist as complex eddy pools in a continuously flowing stream of energy and material elements. That energy from the sun which is caught by life processes is sooner or later radiated away into space, where it is forever lost to us. It takes many millions of years for nutrient elements to be recycled back into the land, and the trickle of energy and nutrients avilable is not infinite and must be conserved. They are conserved in natural systems through closed loops and the webbing of closed loops. That is what food chains are all about—attempts to caputure that radiant energy from the sun and prevent it from being lost as waste heat. The loops of natural systems allow both cycling and self-regulation. Each sub-system regulates its input to release outputs which can inhibit the process that made them.

2. *They release energy in the system in small increments, contributing to homeostasis, the stability in natural systems.* Energy is most efficiently used by organisms in small increments. Mature natural systems consist of many different stages, each of which can store energy for a long time and transport it along a multiplicity of routes and channels. This has the effect of minimizing waste, since, when a sudden surge of potential energy is released in a simple system, very little of it can be stored.

3. *They maintain a steady state, or homeostasis, through negative feedback and permeable boundaries.* The web of electrical distribution is designed as a multiple channel so that power companies can shuttle excess power from one place to another. But we *still* have black-outs in New York and other places. The electrical webs are not quite multiple enough to prevent complete breakdowns. In an integral system, boundaries are permeable and systems mesh together with such intricacy that transactions across the boundaries of systems flow without waste or upheaval. By contrast, our modern urban systems are monocultural crazy quilts, stitched loosely together with waste. We all know what happens when potential energy is released in a simple system and very little of it can be stored, such as when a watershed is denuded of vegetation and can no longer absorb a heavy rain. Downstream flooding results, with a loss of soil and water and a further regression of the natural system. In an ecosystem, many little steps tend to be gentle and subtle, while a few big steps tend to be harsh and destructive.

4. *They store information in a decentralized genetic memory.* Information is the pattern that organizes form. The energy-materials pattern *is* form; and the pattern, when transmitted, is information. In simplest terms, "inform" means to give form to. Within any living system, evolu-

tionary information is contained in the DNA of every cell, and in the learned experience of the food chain. DNA is the code that determines the form and organization of the organism. In human societies, information is stored in cultural patterns, including the built environment, institutional structures and the patterns of communication. It is very important to look at this relationship between information and form, which starts with energy flow. The pattern of energy flow is, in fact, information and culture.

In the continuing cycle of energy materials through time, information is stored as genetic evolution in culture. Information is precious, and needs to be conserved. While species have continuously appeared and disappeared throughout earth history, every major information advance has been retained. The real harvest of the ecological fabric evolving through time is information, including human culture. Returning to a higher degree of regional autonomy or individual responsibility will not be a return to the dark ages, as some have claimed, as long as information is conserved and enhanced. It is a necessary accomodation to the ecological rules by which living beings can maintain themselves in a healthy relationship with their world.

Those points may sound like the jargon of electronics and circuitry, and indeed it is. The language we use to explain the flow of energy and materials in natural systems has been adapted to humanly designed systems. However, the ecological reality is more complex, less mechanistic and more indeterminate than any reality man is capable of constructing.

McCluhan was one of the first people to look at our culture in terms of conserved and enhanced information. We think there is a lot of information in our culture, don't we? We pick up a copy of *T. V. Guide* and think, "Boy, there's a lot of information in there!" Not so. The extent to which we standardize and go for massive solutions is the extent to which we destroy information. Consider the debate about the snail darter stopping the $120 million TVA dam. Now *that* is a parable about information. That snail darter, as a species in a genetic evolutionary chain, contains more information than that damn dam does!

An overload, a surge of energy flowing through a system, can be destructive to its information content. That is a paradigm of modern culture. We have an overload of energy flowing through this system and it is destructive information—it doesn't enhance it. This applies to natural as well as to man-made systems. Run too much current through an appliance and it burns itself out. Overload an aquatic system with nutrients suspended in sewage and its natural health and structure is destroyed. Monocultures process energy and materials so rapidly that the outflow overwhelms systems on the receiving end, and the result is a loss of information and structure.

Economists, those shaman of the modern industrial system, know nothing about the behavior of natural systems, unfortunately. In economics, flow is everything, as measured in money. The destruction of information embedded in the structure of living communities is not measured, and the increased entropy that results from the loss of information is never calculated. Keynesian economics seeks to maximize energy and money flow, while stable and diverse natural and cultural systems are seen as obstacles to progress, as resources to be transformed into cash flow. I see this as burning down the house of life in order to roast marshmallows. This relationship also explains the paradox of inflation without growth. We are developing a system with more and more entropy, but with a greater and greater GNP. There is more flow of money transactions, but no real increase in the quality of life as measured in the structure of information.

The distribution of information in living systems varies with their scale and function, but an analogy can be drawn between the three basic life scales—ecosystems, organisms and cells—and the concept of information can be related to each of these three levels of life.

An ecosystem has five characteristics: it fixes sunlight into a biomass; it cycles nutrients; it is self-regulating in population size; its species succeed each other; and it is self-producing and self-maintaining. The structure of information in the ecosystem follows the general principles that I have already discussed: the presence of negative feedback loops; the existence of boundaries between distinct sub-systems; and information centralized in each cell. An ecosystem can last indefinitely and compensate for major disturbances and dislocations. A lost part or species can usually be compensated for by other parts with a similar function. Information is retained within the genetic, and sometimes learned, patterns of other species.

Man cannot design an ecosystem, and no humanly designed system ever has all of those characteristics of an ecosystem. Gardens and some agricultural systems are comparable to natural systems, and one can design a habitat and whole system—including food and waste—that is built in paths, steps and loops, like an ecosystem. Buildings designed this way will have more stable and healthy systems, but these systems will also be less adaptive to producing the characteristic monocultural surge of quick rise-and-fall cultures.

Organisms are much more centralized with regard to information, especially animals with central nervous systems. They have comparatively little redundancy of functioning parts. We have two kidneys, but only one brain, and the loss of an important organ commonly results in death, although the highly directed and organized parts allows the organism to focus energy very effectively while alive. The organ is the real analogue of the machine; when people understood how organisms worked even at a very elemental level, they were in a position to design machines.

Houses are like bodies, in that their components are designed, like organs, to accomplish specific functions, and you can build in some degree of back-up to compensate for changes and breakdowns. In the house-as-organism, the information is centralized in the humans who operate the household systems. This is another thing that designers tend to forget, and I've had colossal battles with the bureaucracy over this point. I like to think of a window or wall as a dynamic object which people can change, but bureaucratics don't like that. They think that unless a window is triple glazed it's just losing heat. Suppose you put an insulated drape in front of it? Yes, they say, but we can't depend on people to close the drape when it is cold. Do you see where that attitude takes us? The fact that bureaucrats feel they have to protect us from our own stupidity brings us up to a lot of problems. The user has become an anonymous participant, or a non-participant, in the objects and environment designed for him or her.

A cell is the intermediary between ecosystem and organism and design. Its information is relatively centralized in its nucleus, but it has many parallel and redundant metabolism channels, so the cell combines the potential for immortality with the direction of energy toward a specific task. Communities resemble cell tissues in their spatial arrangement and metabolic loping. If you look at slides of cell tissues, they look amazingly like cities from an aerial view. That is not a coincidence. To the extent that communities, or cells, conserve energy or materials effectively, they place less strain on the surrounding and supporting systems.

Farallones Institute is a group of designers, scientists and engineers that I founded about six years ago to begin to put some of these ideas into practice. One of the first things we did was to take an old house in Berkeley and re-design it along some of these principles I am discussing. In the integral house, each major functional system employs multiple pathways from material and energy flow, and Farallones was meant as an example of this. The heating system, for example, includes direct solar gain through windows; a solar air space heating system; and a wood stove space heater for cloudy, cold days. Organic waste can be shunted in a variety of ways. Human fecal matter decomposes in a composting toilet and, when fully decomposed, is used as a soil amendment on ornamental plants. Urine is collected and used as a nitrogen-rich fertilizer. Kitchen scraps are fed to the chickens and converted into edible protein, and the chicken manure is recycled in the garden. Garbage can also be composted or fed into worm cultures to make a nutrient-rich casting for garden use, and the worms can then be fed to the chickens or to the fish in the pond. Duckweed in the pond absorbs toxic fish waste, and in turn can be dried and fed to the chickens. These are all multiple loops. The house is there; all these loops are operating. It's amazing to see the light go on in people as they see these abstractions become real, and as they figure out how to apply an idea to their own homes.

The principle of multiple pathways is closely linked to the idea of diversity and stability of healthy, natural systems, and the multiple pathway is an interactive process within any food or nutrient chain. For example, in a healthy garden, you want to have a diversity of plants, which insures that there is going to be a diversity of insect life associated with those plants. That creates a situation where no particular insect population is likely to get out of control and become a pest. Agriculture is slowly beginning to realize this principle of diversity in crops. The use of pesticides is a postive feed-back cycle: devote one area, the Midwest, to one crop such as corn, and that generates a predator that gets totally out of control. You use pesticides, but the insects become more resistant and you have to apply more and more pesticide. That's a positive feedback cycle.

This idea of integral loops enhancing stability and diversity can be contrasted with its linear equivalent. To heat a house electrically requires the consumption of more times the amount of energy in high-quality fuels than will finally result in useful heat in the home. The methods we use to heat our homes with electricity are like using a nuclear-powered chain saw to cut butter. Richard Stein's book gives many examples of the absurdity of modern design, and the same is true of our methods of waste and garbage disposal. The linear mode we use now simply hauls waste to landfills, and most urban areas are now running out of available landfill areas, not surprisingly. Cities are forced to haul garbage to remote locations hundreds of miles away. Human waste is diluted to a ratio of 100 to 1 with potable water, and then piped to large sewage factories, where the solids are removed with mechanical and bacterial action. The polluted, nutrient-rich effluent which remains is dumped into the river or ocean, where it overloads the ability of natural systems to provide oxygen. This produces a condition known as eutrophication, in which there is a surge of microbiological and algae growth, which then depletes the oxygen and fouls up the water.

Another feature of the multiple pathway is that each component of this system tends to perform overlapping functions, and one test of the integral quality of any system is the extent to which components are integrated into multiple functions. An electric heater can only be an electric heater. A garbage truck can only be a garbage truck. However, a window admits light, provides a view and can also be a solar energy collector. A greenhouse attached to a window can be a solar collector and a storage system or a place to grow seedlings and winter vegetables, or even the location for a hot tub, which we Californians are fond of. A planting box, however small, together with composting buckets, takes the place of the smelly garbage and the noisy garbage truck. Besides processing waste nutrients, it provides a source of food and flowers, and can be the focus for many pleasant leisure hours.

In terms of human purpose, integral design produces value in three spheres: economy, energy use and aesthetics. Economics measures the short-term cost of energy and material transactions in money terms. Costs are set by the availability of and demand for materials and services, as well as by other constraints of the marketplace, such as government actions which affect or distort the economic transactions. The individual decision-maker is concerned with costs that his or her actions can directly affect, and production-oriented economy based on low-priced fossil fuels and so on. That integral design, in itself, very often does not result in short-term economic advantages for its practitioner, is a very important fact. Chances are that if you go out and attempt any of the wonderful things I'm advising now, it may not compete in the marketplace in economic terms. In the short run, a farmer who grows diversified crops and returns waste to the soil may not do as well as the farmer whose profits are gained at the expense of strip mining the soil of its nutrients and quality. The urban dweller who expends the effort to install a waterless recycling toilet still has to pay the tax burden of a wasteful centralized sewage system. The homeowner who installs solar heating equipment still has to underwrite the cost of expensive, conventional, centralized generating equipment, and must compete against the relatively low price of fossil fuels. In each case, the boundaries of the economic system do not mesh with the boundaries of the natural or integral system. The costs of destroying the fertility of the soil are not charged against the soil miner, and there are no economic incentives offered to the individual who does not wish to participate in institutionalized pollution such as practiced in the public sewer.

However, as our way of life runs up against the limits imposed by resource scarcity, breakdowns in its over-extended networks and the simple entropy of bloated size, economics will begin to reflect the advantages of working with the natural grain, rather than against it. The balance is changing, and a future generation of designers that blindly repeats the mistakes of previous generations that grew up without energy as a consideration, will be out of business. If you treat natural systems as a view to be looked at through a thermopane window, there won't be any work for you: work in the future will be in terms of restoring the balance.

In any integrally designed system, total energy flow will be lower than in a linear system, where energy pulses through to achieve maximum product flow—and a high rate of waste. The economic costs are not borne by the immediate user, but get buried in the increasing entropy of larger systems. Nuclear electricity is a good example: the costs are buried in the entire rate base, so you don't know what you are paying for. And a very important point to keep in mind is that a high flow-through does not necessarily imply a high product. Modern agriculture uses 10 to 60 calories of energy to produce 1 calorie of food; it uses more energy to produce than it turns out. In traditional agriculture, that ratio is reversed. In China, it takes 1 calorie of energy—mostly human—to produce 30 calories of food. But there's more to the equation. You also have to look at how much comes out the end of the pipe line. Our agricultural practices are rationalized on the basis that the total productivity of energy-intensive agriculture is higher than traditional methods, and that we would starve without energy-intensive agriculture. Food prices would be astronomical if the current fossil fuel subsidy of agriculture was removed. At a time when our food system is so dependent on energy-wasteful monocultures, that claim is all too sadly true. But more than anything else, it points out the need to rebuild the network of regional energy systems in agriculture before it is too late.

The concept of flow is a very important one. It's not just the ratio of energy used, it's also the concept of the total flow. Integral design tends to reduce the energy flow and use, and integrate more fully into smaller, more decentralized units. For example, we have begun to design food production back into neighborhoods and the home by way of garden vegetables. Even this, as limited as it may be, is very important because it is an energy multiplier.

Aesthetics are too often neglected, but without

them neither economy nor efficiency of energy would have meaning. Indeed, aesthetics is the meaning that we find in form. As conceptual information, it varies from culture to culture: the meaning we derive from the environment is limited by the form of that environment and how we are able to interact with it. A friend of mine from an industrial area of the Midwest commented that while he and his neighbors were surrounded by ecological devastation, they were "tool-literate," because they were surrounded by a forest of tools and machines. Our active participation and interaction with complex and diverse living systems increases our awareness and renews the spirit. In short, the meaning that we derive from life is enriched and constantly recharged by our functional and ritualistic connections to the natural world. If you examine the pre-industrial aesthetic, you'll see that it was always based on that kind of connection, whether it takes the form of a miniature Japanese garden, a simple Arab courtyard fountain or the shade elms of a New England commons. It constituted a whole, picturesque "architecture without architects" of an earlier world that adapted habitat to local materials, site, climate and ecosystem. The important thing about aesthetics is not how something looks, but the kind of participation it brings about with the natural system. That is where the new aesthetic has to be derived from, not from the empty banterings of the architectural establishment.

For the most part, Americans have lost that intimate connection and awareness of their place in the natural world. The "natural" has been reduced to empty symbols, such as the tract house lawn, which originally started as a sheep meadow (the lawn was there because the sheep cropped the grass—but now the sheep have been replaced by lawn mowers). Ecosystems have become dim landscapes to be appreciated out of a car window, another low-information object like the T. V. screen.

The task, then, of integral design, is to begin to recreate the opportunities for people to derive meaning and satisfaction from their experience with natural cycles. This assumes that people become active and intelligent participants in managing and maintaining their environment. The hotrod is an example of an aesthetic that grew out of our attempt to find meaning in everyday industrial culture. Maybe the day is not too far off when millions of Americans will be hot-rodding their now denatured houses into finely tuned, multi-channel, closed-loop, organic instruments for processing nature's flow.

4. Architecture in General / Architecture, Climate and Energy in Particular

Ralph Erskine

Moving to Sweden was for me the source of many strong impressions. Only one of these, but a potent one, was the impact of the climate, the light summers and the long, dark, cold winters, the need for protection and effective heating and lighting that the winters created. Climatic and energy-economic architecture were major interests of mine for many years. Today "the Arabs" raise the price of oil and the cost of energy begins to approach its value, and each time the OPEC ministers meet I find more of my earlier form-and-construction theories becoming financially, as well as economically, justified.

Having observed the fashions and results of the "Architectural Rat Race," I fear that low-energy architecture may become a new "ism" as we move our interest from one to another of the many detail aspects of architecture. I intend, therefore, first to discuss the general perspective, so that we might bear it in mind as we concentrate on the special subject of this meeting, Hopefully, as yet another physical attribute and new "forms" are added to the language of architectural "objects," the "subjects" of architecture—the users and the community—will not be forgotten.

Is it possible that we may be aided in this by the fact that more and more women are becoming involved in architecture? A third of those in our office are women, and I discern that they may have an impact on architecture which in some ways will be more important than that of men. Without wishing to generalize, I would observe a certain truth in the old truism that there is a tendency for men to be more interested in things (such as ARCHITECTURE!) and for women to be more interested in people. A great advantage, if it is an "architecture for the people"—which is needed!

I consider being interested in people to be the first prerequisite for the creator of good architecture, but there are many other criteria. Do our buildings and towns fulfill all our reasonable everyday needs, but also point the way to some better human community? Do they encourage contacts, but also give personal privacy? Can they inspire those who use them to partake in the task of giving them form? Can they give the buildings life and, with understanding, use them in expected and unexpected ways? Does our architecture contribute positively to heightening the value of what previous generations have created, and does it open possiblities for future generations to adapt it to changed needs? Architects know of world poverty and the shortage of resources. Is, therefore, their architecture economical, and does it achieve the desired goals with a minimum of physical and economic means? Does it tend to create satisfying work for those who build? Is it amenable to the organizational, financial and production processes which exist, or—should these make it impossible to achieve the good aims of the users and the culture—does it contribute to the changing of hindersome conventions and outdated practices: Can it, as all created art should, disturb our complacency, but also give new and unexpected pleasure without overriding the above criteria?

I fear that it is seldom that *all* these and other important criteria are affirmed in our work, and too often I am left with the feeling that our architecture belies much of what we and our culture proudly declare to be our highest ideals. Sadly enough this is most apparent in our capital cities, our universities, and our most representative buildings in the very centers of our culture. And sadly enough it is often also most apparent when the symbolic and aesthetic ambitions of the architect and client are most intensive.

Traveling to Cornell I pass through London or Paris, and New York—three important cultural centers. Of these, New York would probably be considered the most vital center of modern, western culture, and the most beautiful and awe-inspiring example of modern city building. It is a city which makes a tremendous impact on me each time I visit it.

The fantasy of its silhouette, the audacity of the skyscrapers, the suave beauty and smooth technology of the long generations of simple glass pillars from "Lever Brothers" to "Citicorp," (an architecture of disciplined "Saville Row Aristocracy" which now may be broken by Super Tallboys and other affectations)—here in Manhattan is a townscape of great contrast and great formal beauty. The language it speaks is clearly of "modern high technology," but what else does it say, and what does it symbolize?

A most apparent symbol is the passage of New York Avenue through alternating areas of great luxury and privilege, and ghettos of the poor and discriminated against. The aristocratic or ecclesiastical potentates have been superceded by the Committees and Boards of our corporate civilization, but the luxurious glass "palaces" and dilapidated slums are recognizable as a true representation of a Medieval, a Renaissance or perhaps a nineteenth century society without cognizance of the world about which we so proudly preach!

The exotic and expensive buildings of Oxford or Cambridge, of Harvard or Yale, demonstrate the true belief in exclusive privilege of our universities. Regrettably this is, to varying degrees, a characteristic of architecture in most countries of the world, irrespective of their race, religion or political system. It is a true world culture! The denizens of the Middle Ages and Renaissance knew that equality of rights was neither sought by Man nor ordained by God, and their architecture was an honest and beautiful expression of this belief. Is the "modern" architecture of the democratic states the most dishonest in human history? Can this be a greater problem for our time than the inefficient energy balance of our build-

WARM
GREEN
SPRING
SUMMER
NOON
24 HOURS DAY
NOON
JUNE
JULY
MAY
AUG
APR
SEP
MAR
OCT
FEB
NOV
JAN
DEC
24 HOURS NIGHT
COLD
SNOW
ICE
STERILITY
WINTER
AUTUMN
R ERSKINE
THE RYTHM OF THE SEASONS
THE PULSATION OF LIFE

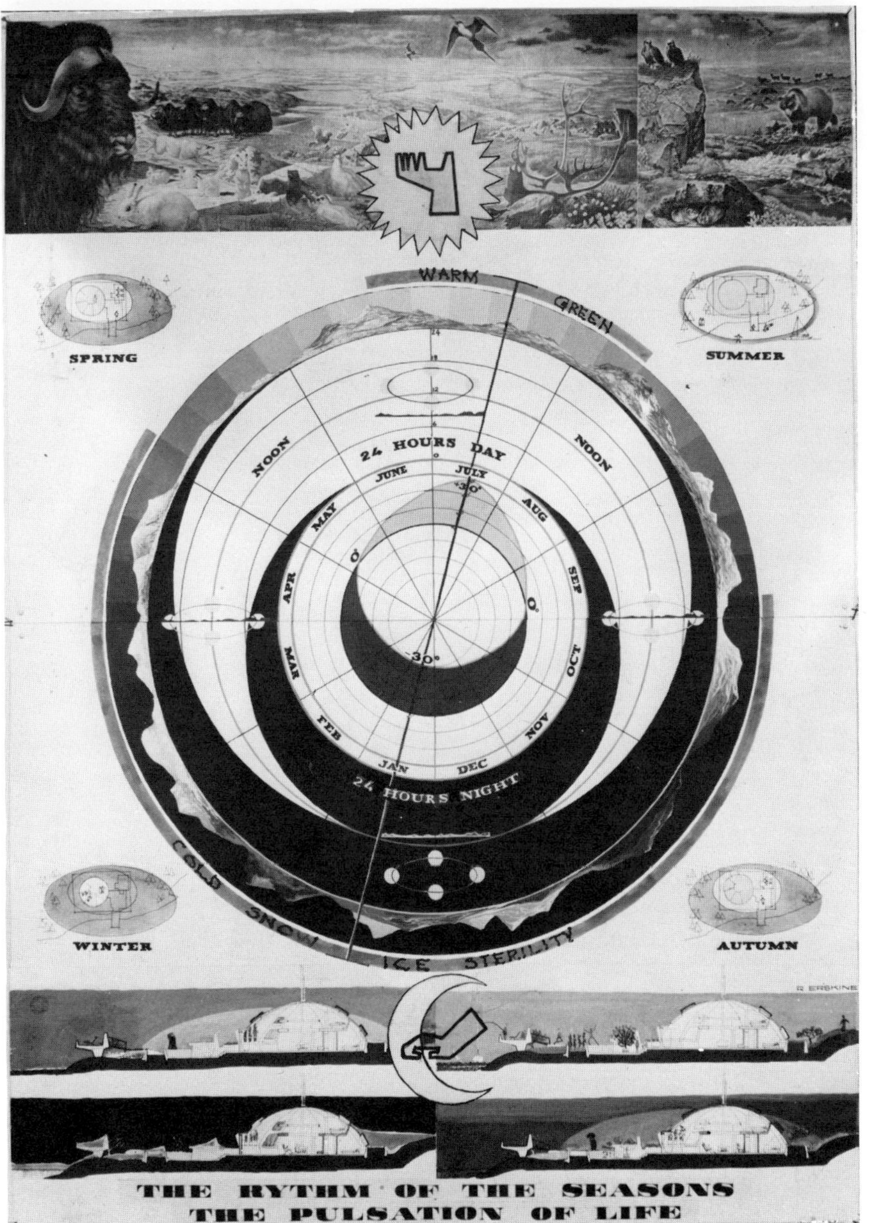

4-1

ings? Would not worldwide human justice also mean that economy in the use of *all* resources would become a "must" and energy-economy an important detail characteristic?

Architecture, Climate and Energy

In my early work in Sweden I talked about the northern situation and made a conscious attempt to fit into it. This I obviously had to attempt since I wasn't indigenous to the North. In Sweden, my chosen home, it is of course cold in the winter and warm in the summer (Figure 4–1). For the Swedes, Finns and Laps that is obvious and natural, and they have many traditional techniques for meeting the situation.

At mid-winter in the north, the sun does not rise above the horizon; in mid-summer, it is eternally light. The climate changes, and with it the landscape changes from the cold sterility of winter to the teeming life of summer; even life itself changes to a fantastic degree among all the people who are used to living in the situation. In the winter they largely live within a well-protected dwelling, only making sorties for work or sport. As spring comes they move out into the protected near-environment for relaxation; with summer, into the surrounding landscape, often deserting the permanent dwelling entirely. With autumn they move back into the near-zone and the protective interior. These are very important characteristics of the spiritual experience of the life-cycle of the north.

There are certain very discernible parallels between the cold and the hot, dry climates. While the spiritual experiences of living in the north or living in a hot, dry climate are extremely different, the physical characteristics of heat movement, of snow or sand drifting, of vegetation, are in many ways extraordinarily similar. For buildings themselves it makes very little difference whether the heat is outside or inside. Techniques for meeting these situations could be surprisingly similar. They often however, are not, since the respective experiences of the "north" and "south" don't transfer well, and very few people can move between the two.

The Cold

In trying to come to terms with my new living and work situation, I made a series of simple analyses. A cube or sphere has maximum volume with minimum surface; since heat loss or gain happens through their perimeter, buildings would approximate these forms in extreme climates. The separation between indoors and outdoors is an essential condition for civilized life, and this would be emphasized for both technical and artistic reasons. The ground is protective since temperature variations there are not as extreme as in the air. A semi-sphere could therefore be a suitable building form for the north, while a sphere would be patently unsuitable, since it would expose its underside unnecessarily.

After constructing a protective envelope and arranging for entry and exit through a system of air locks so that uncontrolled heat and air movement is avoided, we have conquered the traditional enemy of cold (or heat). People may then overreact, and one can find tropical heat in the houses of the north and subarctic cold in the air-conditioned tropics.

The Seasonal Change

As warmth comes to the north we desert the man-made surroundings of the house, as mentioned earlier, and go out to nature—to the islands, the lakes, the forests and the hills with which the North abounds. We live simply and enjoy the short summer while it is there. When I first arrived in Sweden I could not understand why a country with so short a summer had so many boats, week-end cottages, open-air dance floors, theatres, swimming pools and restaurants. It was only after some years of direct experience that I realized it was precisely because the summer was so short that we needed these facilities. It is a precious time of hope and joy for which you long. Also, this is part of the impact of the seasonal rhythm in which we live!

In the autumn one moves back to the city or the nearness of the home, to the protective wind-breaks and sun-traps which must bc arranged in the surroundings. In the winter one retires to the fortified interior, with its tropical freedom within the protective shell. Moderate-sized windows are placed to give light and enjoyment of the view, and nature is experienced through the windows, with occasional excursions outside for different activities. (Figures 4–5a, 4–5b, 4–5c, and 4–5d)

That is the generative base of architecture in the North, but I did not find that it had formed the architecture of Alaska or of the Canadian North, and you won't find it very consciously created in Scandinavia. There, it is not architecture but craft—they just do it.

Snow

Snow is a gift that indigenous people have always used. The Eskimos use snow for construction, and Scandinavian peasants—and I'm sure farmers in the American prairies and in Canada—pile snow up against the house to improve its insulation, leaving holes for the windows. In the dark of the long winters the northerners long for the snow to come to ease their movement through the country and give reflections of a lighter, gayer landscape.

A ski hotel I built in Borgafjäll uses those principles and integrates the building into the surrounding landscape. The snow and ground protect and insulate the building, and the patterns the snow makes on the ground create a large, natural sculpture that can be enjoyed by the guests, both visually and as practice slopes (Figure 4–2). Upon reflection, this seemed to me a truer and more sophisticated technique and aesthetic than my original sketches for "a Glass House in the Subarctic."

Outdoor Protection

It is natural that Swedish architecture is not always responsive to our climate. We have a beautiful building, the Stockholm Town Hall, built with lovely arcades that you associate directly with those of the Doge's palace in Venice. But Stockholm is not Venice. Most of the year you walk across the Piazza San Marco and the intense heat

4-2

burns your head and feet. With joy you move to the arcades which surround the square and to the delightful shade and cooling breeze. In Stockholm as you move from the open space to the arcade, exactly the opposite happens. You move out of the warm, welcome sun and into the cold shade and draft. The forms and the architectural instruments are identical, but the instruments were appropriate in one situation and not the other; and the vocabulary that was invented in Venice worked for Venice, but not for Stockholm. The Italians quite rightly ignored the longest period of our yearly cycle—the winter—and adapted to the shortest. Instead of imitating the attitudes, understanding and techniques of the Italians and Arabs, we have imported only the architectural forms, forgetting the particular situation of which they are a part.

I do not consider it to be of major importance whether architectural form is "organic," "new rationalist," "brutalist," or otherwise. It is, however, of the utmost importance that architecture be the art of the useful. It gives a special "joy in use" with the realization that its special beauty stems from the wise and subtle way in which it is so well adapted to satisfying our many needs. This can be illustrated by the separation of outside elements from the structural envelope. By not breaking the structural envelope, cold cannot be conducted from the outside to the inside. For example, when a floor is extended to form a balcony. (Figure 4–9)

Could a new architecture be invented which would catch our sun and protect us against our chilling breezes, which in *our* situation would create those comfortable conditions of temperature which are identical for Swedes or Italians? If a northern architecture could do this as well and beautifully as the arcades of Venice or Perugia, it would certainly have a very different form.

Vegetation

Vegetation, in Sweden, is sparse and often needs to be protected. In Versailles they wished for orange trees—which are ill-adapted to a North-European climate—so they moved them indoors for the winter and enjoyed them in glasshouses. In the whole of Scandinavia you take many kinds of plants and animals and birds with you indoors for the long winter period. In Kiruna in the subarctic, they have no permanent outdoor gardens. They put out their potted flowers for the few days or weeks that are frost-free, and then take them back indoors as the autumn approaches.

The scale of vegetation must also be taken into consideration. I was used to the luxurious growth of England, that warm, rainy climate which is so wonderful for vegetation and flowers. The English arrange flowers in enormous bouquets. In Sweden flowers are more delicate and sparse, and it is perhaps because of this that Swedes arrange flowers sparingly and talk about them as individuals—much as do the Japanese. It may well be that the varied way flowers exist in nature is a useful analogy for the difference between these two cultures. Could the same sensitivity be found in our building or landscape design?

The courtyard of my house in Drottningholm has many hard surfaces. There is a dual reason for this. I mentioned earlier that in the summer the Swedes use the whole landscape, that they leave the cities and move to their summer homes. A large percentage of the population has summer cabins. This means that for us, as for others, the permanent garden gets left behind, but our concrete and granite is maintenance-free during the summer and can be quickly complemented with flowers upon our return in the fall. In spring and autumn when the air is cool, the sun heats up the stone and improves the microclimate of the court in the same way that grass and trees and fountains tend to have a cooling effect in a tropical climate.

Aggregation

In certain situations, a multitude of functions can occur together within one common skin. This was the idea behind the Lulea Shopping Center where the street continues inside the building, protected and heated but supplied with openable windows for those occasions when the outside air is sufficiently warm. (Figure 4-3).

There is a whole rhythm of "passage" of which one traditionally feels a part during the northern summer, when the sky is light all night. But in the winter all the social action moves indoors and becomes private or semi-private. I thought of the Arabs, with their mats over the streets, and of the Italian arcades, and how beautifully they had protected their social spaces against the stresses of their hot summer. Could we not equally well protect against our cold?

I therefore extended the main street of Lulea into the heated interior of our building, and the world's first indoor shopping center was built. We included in the center a big department store along with many smaller stores, cafés, a cinema, restaurant, exhibition rooms, streets, alleys, and a central square for music, dances, speech making and entertainment. It became a city within a city and a useful complement to the outdoor streets and squares.

Those social activities that go on outdoors also take place indoors, but they are experienced very differently. In indoor environment the consideration of scale and its influence on the character of social contact become very important. Should the scale created not be large enough it may be "private" rather than "public" places which arise. If it is too large or anonymous the social exchange can become aggressive.

4–3

Microclimate

Microclimate in the North is extremely important. Avoid the breeze and trap the sun is a basic rule which would apply during most of the year. Due to the low angle and 24-hour duration of sunshine, solar gain can be an important energy source, but at other times it becomes a major problem. Surprisingly perhaps solar control can be of considerable importance in the high North, and different forms of variable windows would be of great value.

A township in the north of Sweden is designed around these principles: a south slope, a view of the landscape, higher buildings on the north edge where they will never shadow other buildings, a view of the mine workings and of the road so that you can enjoy traffic movement and all the threads of contact with the outside world in this isolated situation.

A sketch I earlier made for a township in the North looks like an Arab town. Its vivid colors of brown, yellow and red provided a positive experience in the cold, colorless winter landscape. On the whole, small windows were used; large ones only occasionally in especially important places (Figure 4-4). This is obviously the "low technology" way of coping with either a desert or an arctic situation. In both cases there is isolation, a great need for protection, and the small amount of vegetation has to be carefully protected and nurtured. It was this sketch which for me gave rise to the thought that there could be considerable similarity between the cold and the hot-dry climates, a thought that I have later pursued further.

One problem which I realized when regarding my sketch for the township in the North was that it revealed a one-situation town. Everyone lives within the town wall.

Walled towns must have been very claustrophobic, and nearly all medieval walls were pulled down as soon as weapon technology or political organization made the walls unnecessary. The stones were then sensibly reused—recycled one would say today!—for building other useful buildings, and the barrier between those inside and outside the wall was broken.

In later projects I have therefore modified the screening building which created this introspective situation for all the different types of buildings, for the separate or row-houses, the flats, schools, workplaces, church, shopping centers, etc. Especially in semi-desert situations, nearness to others and to service facilities can be very positive. There are, however, many people who like the convenience, but not the emotional experience of this nearness; and since we are hopefully designing for a relatively free society, we obviously should offer as many choices as possible. Privacy needs to be protected, for while visual contact and the feeling of being with others can be of great importance for many people and much of the time in these isolated communities, most people, some of the time, and some people, most of the time, could suffer from over-exposure to social contacts. The form of my township and the protective wall changed to meet these insights.

When planning for Eskimos and southerners in the Canadian Arctic at Resolute Bay, I realized

SYNTHESIS

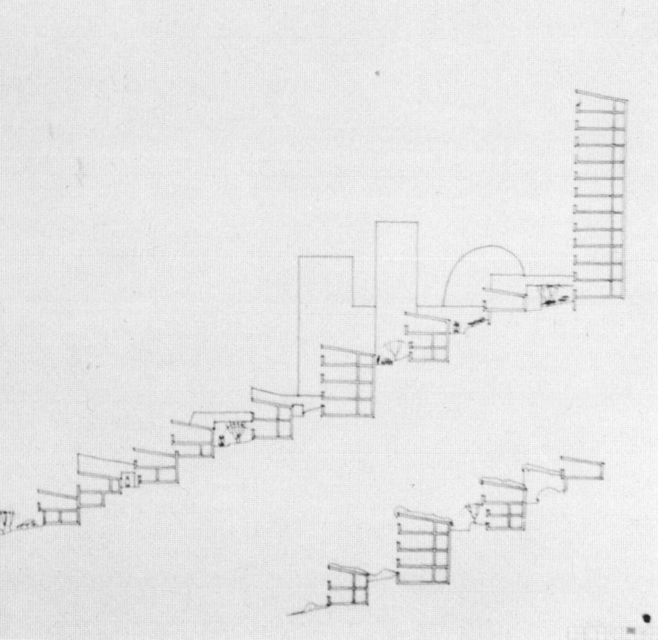

4-4

that many such modifications might be necessary. There in the High Arctic, physical problems are great but the human and social problems can be greater. Here, as everywhere, technical solutions must be subservient to subtle human needs. Science fiction projects for covering whole townships are naive and inimical, and I even found myself designing many one-family houses for Eskimos in this extreme situation so that they more easily in their isolated township might establish their family identity and their relationships to nature, to one another and to the southerners of another culture who would be their neighbors.

Since humanity in the eighteenth century discarded the medieval realm of mysticism and developed a scientific system for studying the physical world, the results have been sensational–insufficient in some ways, but nonetheless sensational. Humanity has more recently commenced the careful study of us human beings. This continuous and careful search and painstaking checking of results will, with time, lead to equally sensational realizations, results of which will be of vital importance for architecture and community planning.

In the meantime, we architects must seek such knowledge wherever it is to be found. We must use it and our intuition and sensibilities and experience of people to fulfill real and often prosaic needs with a poetry which illuminates these important realities, which tell of the dreams of justice and equality.

To make a difference between what I would term "modern" architecture and "new renaissance" architecture which is so prevalent now, we must work for political, economic and administrative instruments which will ease our task of making reality of these dreams. Our hope is for relevant change; our allegiance is, therefore, with radical rather than conservative philosophy and politics, and with the very real needs of the needy and underprivileged, rather than with the profitable commissions and dubious needs of those who are powerful and rich.

Figures 4-5a–4-5d: This domed house demonstrates the three main areas in a Swedish home, and how those areas relate directly to the Swedish climate and seasons. The window area is a focus of activity in the winter; the outdoors area is well utilized for summer activity; the semi-enclosed intermediate area with a warmed sunscreen is in tune with the needs of autumn and spring living.

4–5a

4-5b

4-5c

4-5d

South elevations to catch the sun, and small north elevations to avoid long shadows are the reasoning for the roof in this design which is one of the main features. 4–6

4-7

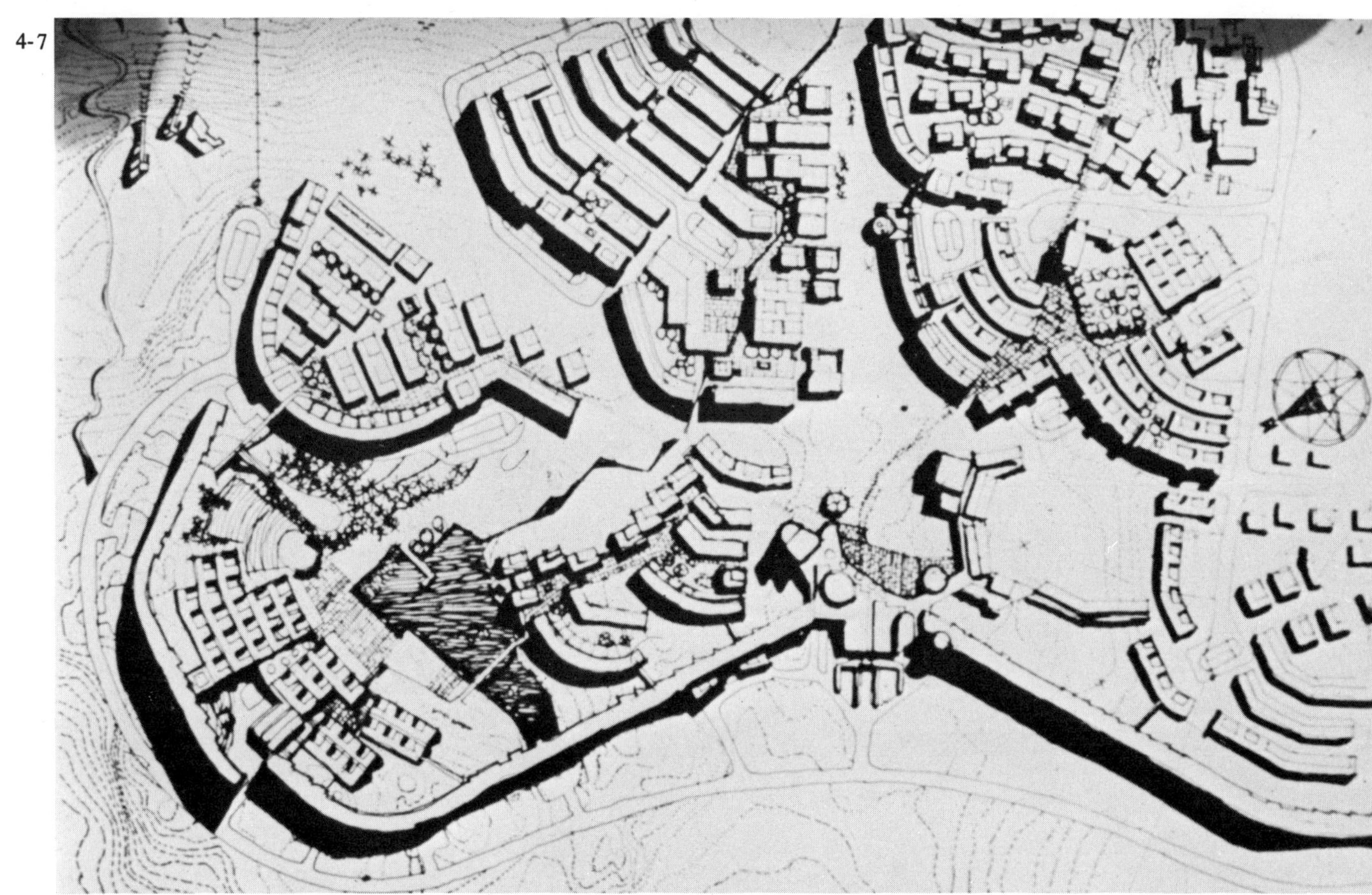

Illustrations of arctic town project (1958). The town turns away from the northerly wind and towards the sun. A south orientation is necessary to provide the maximum amount of sunshine obtainable when radiation is from a low angle. The site is chosen for its favorable location.

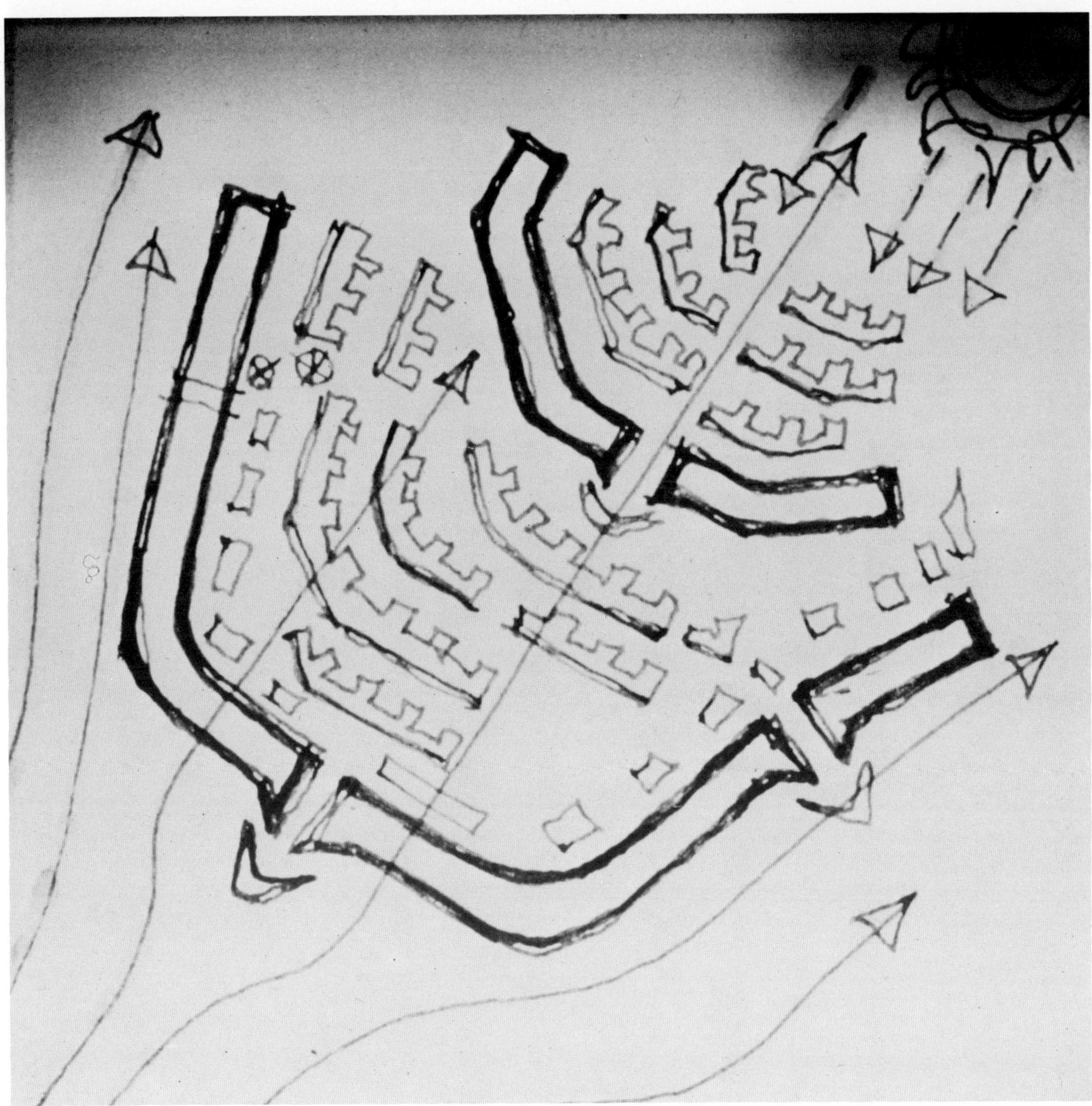

The wind usually has a cooling effect and, therefore, shelter from it is important.

4-8

Structural separation and protection have both been initiated by considerations of climate. It is important, however, not to break the structural envelope, which could result in cold-bridges and danger from frost. For example, extending an inside floor into a balcony creates a conductor for the cold to move into the interior—that is what I mean by a cold-bridge.

4-9

Figures 4-10, 4-11 & 4-12: It is very important to separate outside elements, such as hanging balconies. The break prevents the heat loss, or cold gain, that would otherwise occur. I consider this of such great importance that I also emphasize it visually.

4-10

4-11

4-12

5. Solar Access, Rhythm and Design

Ralph Knowles

I have recently been involved in conversations that have led me to re-evaluate some of the comments I've made in the past and to cast them in somewhat different terms. In general, our concern for a number of years now has been overdrawn in relation to energy. A lot of that comes from years of having to overstate the problem in order to clarify our own thinking. We exaggerated. I'm guilty of it, as are many others. We tended, I think, to overstate our hypotheses, to oversimplify the facts and thus to make very important things somehow trivial in our attempts to describe them. While our intent was courageous, noble and just, we tended over time to deal with the issues in a manner similar to the way lawyers attend to the law instead of to justice.

I would lay claim to some guilt, for example, in synthesizing a somewhat wordy hypothesis from some writings of the 1940's and 1950's. In *America Building,* Fitch stated a very simple case of overreliance on energy-intensive modes of adaptation, or mechanical energy in place of location and form as design adaptations. We were substituting things for other things. This single paragraph had a tremendous impact on me and I combined it with a couple of other simple statements by other people. Kevin Lynch, for example, wrote a very influential book called *Image of the City,* in which he worried about the non-structured urban scene. A. E. Parr was a third influence. He talks eloquently about our dependence on a sensory system that is evolved out of unimagined time.

These three sources together led me to, I suppose, an overdrawn hypothesis. I tried to state it in simple terms: "An artificial system made in balanced energy response to nature will exhibit diversity useful in expanding our choices, therefore improving life quality." In part, it was a notion founded on the basic structure of our perceptions: that, if the environment were not diversified, there would in fact be no choice.

Later on, as I attempted to test and expand my hypothesis, I learned about the Piute Indians of the Owens Valley in California, who were already making energy choices for life quality. The Owens Valley is 100 miles long and 20 miles wide. It's bound on the west by the Sierra Nevada Mountains, 14,000 feet high, and on the east by a lower mountain range, much softer and much older.

The Piutes had their permanent settlements in the foothills of the escarpment of the Sierra Nevada Mountains. They stayed in fairly permanent stone shelters during the winter. As the summer approached, they migrated westward, up into the

Sierra Nevada Mountains, where the atmosphere is cooler at the rate of about 3.5°F for every 1000 feet of rise. The Piutes were able to control their climate by migrating.

After spending the summer in the high Sierras, the Piutes would migrate down off the mountains, across the valley floor and into the softer Inyo-White Mountains. Here, the slopes faced south and southwest. The Piutes would harvest pine nuts there and stay through fall and early winter. As deep winter approached, they migrated back to their permanent stone shelters. Then the cycle would begin again.

At particular places along the way, the Piutes would do particular things at particular times. One can lay a grid over the cross-section of the valley and mark not only places and time on it, but events.

It's interesting that the Piutes always migrated east–west and never north/south. To migrate north–south was easier. If you get down on the valley floor, it's a straight-line shot north and south. But moving in this easier, straighter line would not have increased their choices for life quality. If they had stayed on the north–south path, they would have stayed within the same ecological condition, so that they could have traveled long distances and never have improved or diversified their choices. It was in the east–west direction that they could increase choice, moving from one microclimate to another.

Talking with the other contributors to this symposium, about automobiles and the freedom of choice the automobile provides or doesn't provide, I began to think about the Piutes, and it seemed to me that one can make a kind of distinction between good energy expenditures and bad energy expenditures. My intent is not to condemn all energy expenditures, but to distinguish between those that increase choice and those that do not. In Sim van der Ryn's words, between "those that increase information" and those that do not. Consequently, to the extent that variety and diversity allow choice, those expenditures of energy that increase choice can be an element of life quality. And life quality is my concern.

This is somehow a clearer design idea; it is closer to the designer's intent to think about life quality than to think about energy. The notion of quality, after all, is almost by definition the designer's domain.

In order to increase choice, I suppose that we are talking about choice to some generalized figure we call the user. But before the user's choice can be increased, there are a couple of characters that come between. In our society, they are the whole developmental mechanism, the developer and the whole design mechanism. In my case, I'll talk only about the architect. The architect's choices have to be increased. It is in this regard that I want to talk specifically about solar access, rhythm and design. Without access to the sun, a whole range of design choices are not available to us. A whole range of development choices are not available to us. Finally, a whole range of user choices are not available to us.

I see solar access in terms of a new aesthetic: a design strategy based on rhythm. The Piute's lives demonstrate a natural tempo, or rhythm, sympathetic to the recurrences of nature. The migrations of the sun from east to west on a daily basis, and from south to north on a seasonal basis, may also be considered rhythmically, with important implications for architecture.

If we accept Simon's notion of design as "dealing with the world as it ought to be," we are left with the question of new images and how to develop them. The instances of solar adaptation in the natural world and their promise for life quality provide new models to inform our imagination. Of course, consideration of solar access is not a new aesthetic of itself, only the solution to a problem. But the potentials of solar access as an element of life quality, of time as an element of design and of rhythm as a design strategy present significant new choices to us as designers. The thread I would like to carry through here is a design thread, focusing primarily on life quality, and not primarily on the conversion of energy. Two photographs illustrate these concerns. Both are of buildings, or groups of buildings, that I regard from a number of points of view, but will use to prove a point.

In Figure 5-1, two buildings sited in relation to one another (for probably good reasons) reveal a relationship we don't often see, simply because we are not concerned with it. One building shadows the other, with energy consequences. The adaptive systems are mechanical. Conditions defined by the shadowing will vary; the loads will differ; and the mechanical systems will have to respond. There is great energy cost in that response; there is a control cost in that response. It is a situation that is not normally taken into account. It's not even a situation that we see in other than sculptural terms.

5-1

5–2

Figure 5–2 has less to do with energy conversion and much more to do with an obvious aspect of life quality. This is a housing arrangement in the downtown part of Los Angeles, in which for good and appropriate reasons buildings are grouped around a public space. But here the public space is a recreation area where people can swim and sunbathe in Los Angeles during the time of year pictured. And the project is shading itself, thus removing options.

Now I would like to go back and pick up the thread of the North American native, because there is an issue of life quality here that I think is of great consequence. And it is related to the conversion of solar energy. Figure 5-3 shows one of the sets of buildings, Longhouse Pueblo, at Mesa Verde, Colorado. It's a huge cliff dwelling that faces out to the south; this is the west end of it. It is a terraced arrangement, stepping back into a cave that's over 500 feet across, 150 feet high and about 130 feet deep. The construction is of stone, which was a native material.

5–3

Moving back up into the cave (Figure 5–4), we can see an elegant fitting of site and building. In fact, it is difficult to tell what is building and what is site, or what is built on the site or over it. The form of the cave also allowed water to be collected from the natural dripping action; this not only provided domestic water, but helped provide cooling as well.

5–4

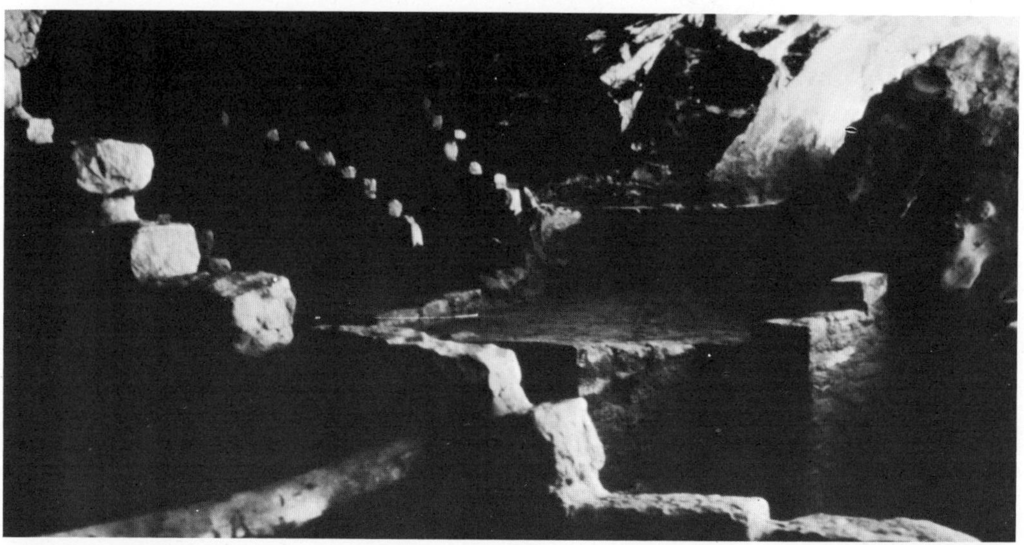

5–5

In a primordial sense, the cave is shelter (Figure 5–5). But its visual richness alone would stun any architect. There are contrasts of light and dark, hot and cool, white through a range of colors. The textures of the stone read as tapestries. The east end of the cave (Figure 5-6) bears water markings and smoke markings.

5–6

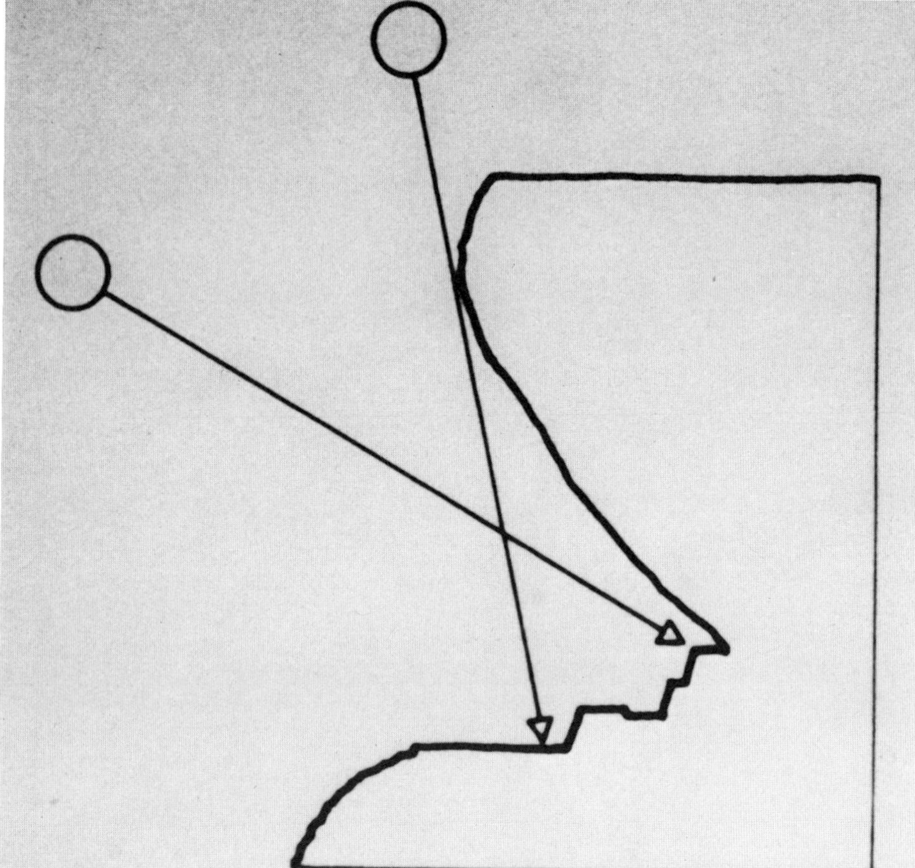

5–7

The cave functions as it does because of its south orientation (Figure 5–7). The low winter sun comes in, heats and lights all of these surfaces. The heat is stored in the mass of the cave. The summer sun is higher, but the great cyclopian brow of the cave protects most of the cave surfaces and buildings and keeps them in cool shadow. Figures 5–8 and 5–9 illustrate the seasonal rhythm of winter sun and summer shade on the pueblo dwellings.

A composite plan of all the layers of the terraces (Figure 5–10), starting at zero feet, up to 3 feet, 5 feet and so on, makes clear that the terraces are not only an effective energy system, but a clear kind of spatial system in which it is very difficult to get lost. The system provides for orientation, not just in space as Lynch wrote about, but in time.

The winter sun penetrates deeply, and as the sun migrates by season to the north, the shadow moves to the south. As the fall approaches, the reverse progression occurs. In the spring, it reverses again, so there is a kind of in/out migration of the people's work, by season. That much we know for a fact.

I began to speculate on the possibility that there were diurnal sorts of migrations, or pulses of activity, as well. For example, the morning sun comes in, travels across the face of the cave and ends up on the opposite end. Now we can play a kind of scenario: The morning sun comes in and wakes up the turkeys, which the people kept; the turkeys gobble and wake up the children; the children wake up the parents, so that the first pulse of daily life begins at the west end of the cave and echoes 500 feet across to the other end of the cave. In the afternoon, the reverse happens. The whole west end of the cave goes into twilight,

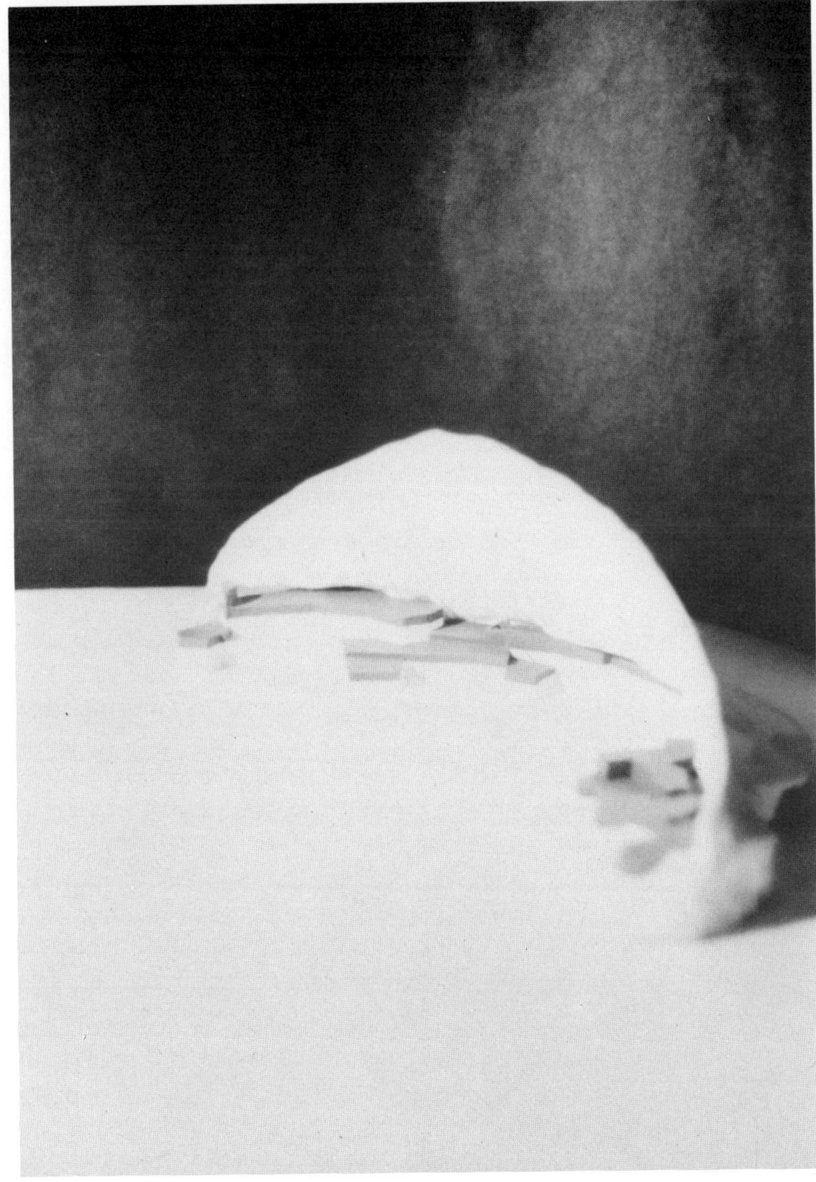

5–8

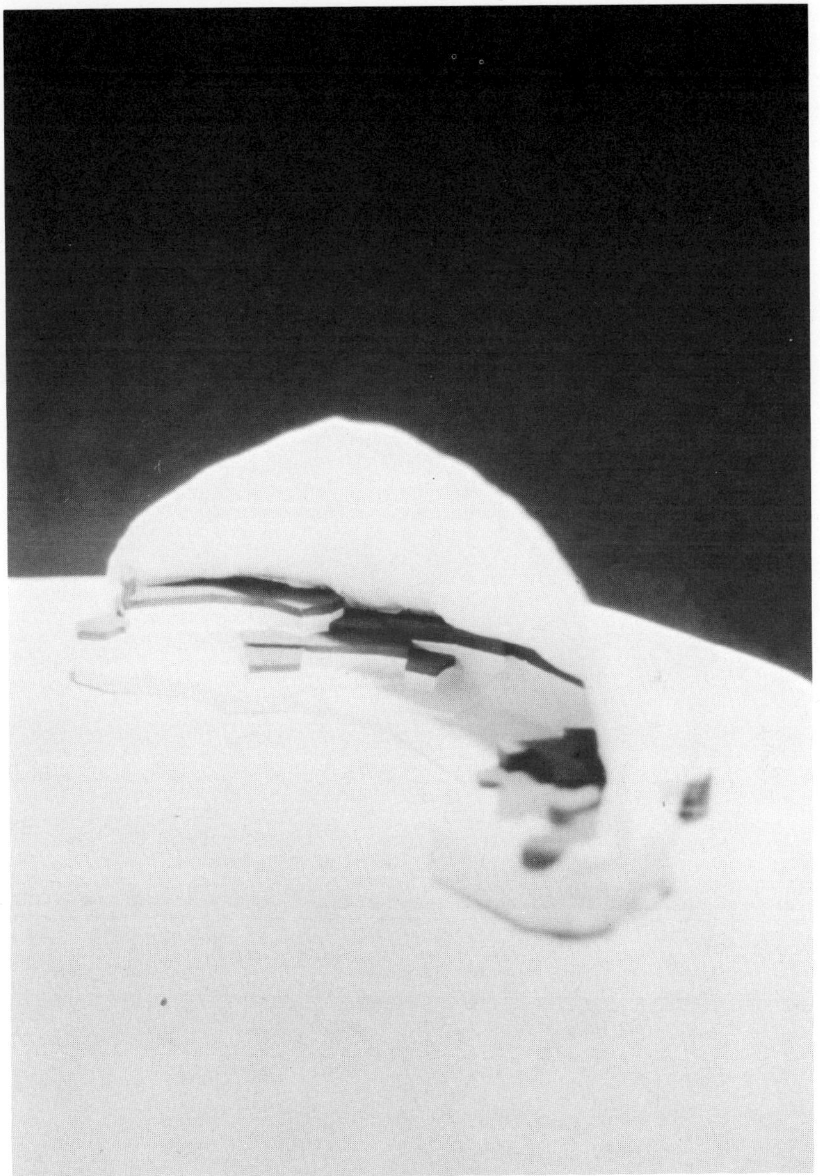

5–9

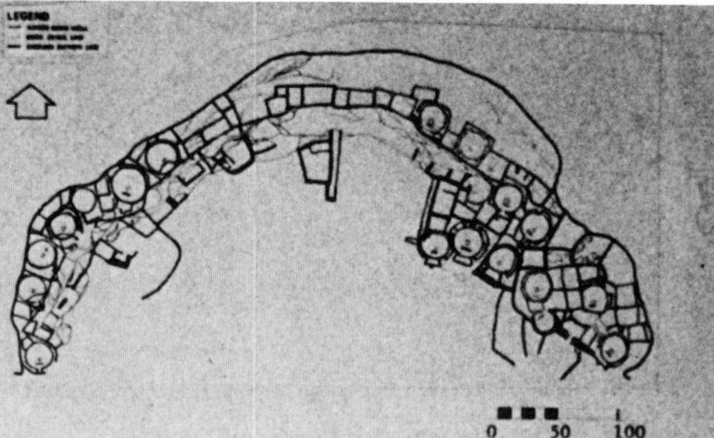

Diag. 9: Plan of Long House, Mesa Verde; based upon a reconstruction by Dr. Douglas Osborne, Prof. of Anthropology, California State College, Long Beach.

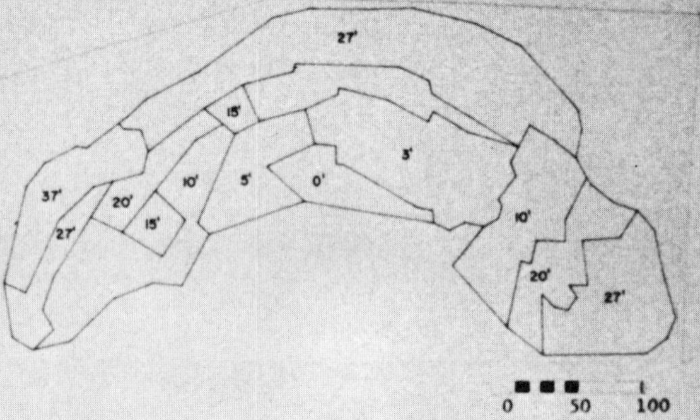

Diag. 10: Contours of the Longhouse plan showing levels of major terraces and building groups; the elevation numbers relate to an arbitrarily selected reference (0') at the cave front.

Plan Drawings: S.Jhono, R.MacDonald, & R.McMahon; 1967

5–10

while the evening rays of the sun are still penetrating on the east. So that while the turkeys on the west end of the cave are starting to quiet down and the children begin to nod and the parents start to breathe that last daily sigh of relief, the parents at the east end of the cave still have a long way to go.

The annual scenario, which we know, and the daily scenario, which is conjecture, illustrate a counterpoint of rhythm. I can link the sense of rhythm at Longhouse to the rhythms of another native people, the Piute Indians, as they migrated annually from one side of the valley to another side of the valley. I can even see in these rhythms the possibility of a design strategy, in which the demarcations of time can be manipulated for design purposes.

Figure 5–11 shows Acoma, another pueblo, located about 50 miles from Albuquerque, New Mexico, on the top of a plateau. Looking at it from the south, one sees it is a set of row houses, separated by streets. The plateau is a very small one. They didn't have much space, so they didn't farm on it, but farmed the land below. We'll come back to this point later, but if you look closely, you'll see that the spacing between those rows of buildings varies. It also happens that the lengths of the shadows vary similarly, so that there is a corre-

5–11

5–12

lation between the size of the shadow and the width of the street. Figure 5–12 shows that Acoma is a terraced arrangement, like Longhouse.

Figure 5–13 is a standard cross-section through the house. It has three levels (corresponding, perhaps, to the belief that there are three steps to heaven). The lowest level, which has the highest ceiling, is for storage. The residents stored their food down there, where it could maintain a relatively steady thermal state. The storage area was large because the farming area was some distance below the pueblo itself. This arrangement not only provided sufficient space, but enjoyed a low surface to contained volume ratio. The people slept at the second level and cooked above that, on the third level, where the fire and smoke could rise and not get in their eyes. The terraces face south and have a high surface to volume ratio, while the north has a low surface to volume ratio.

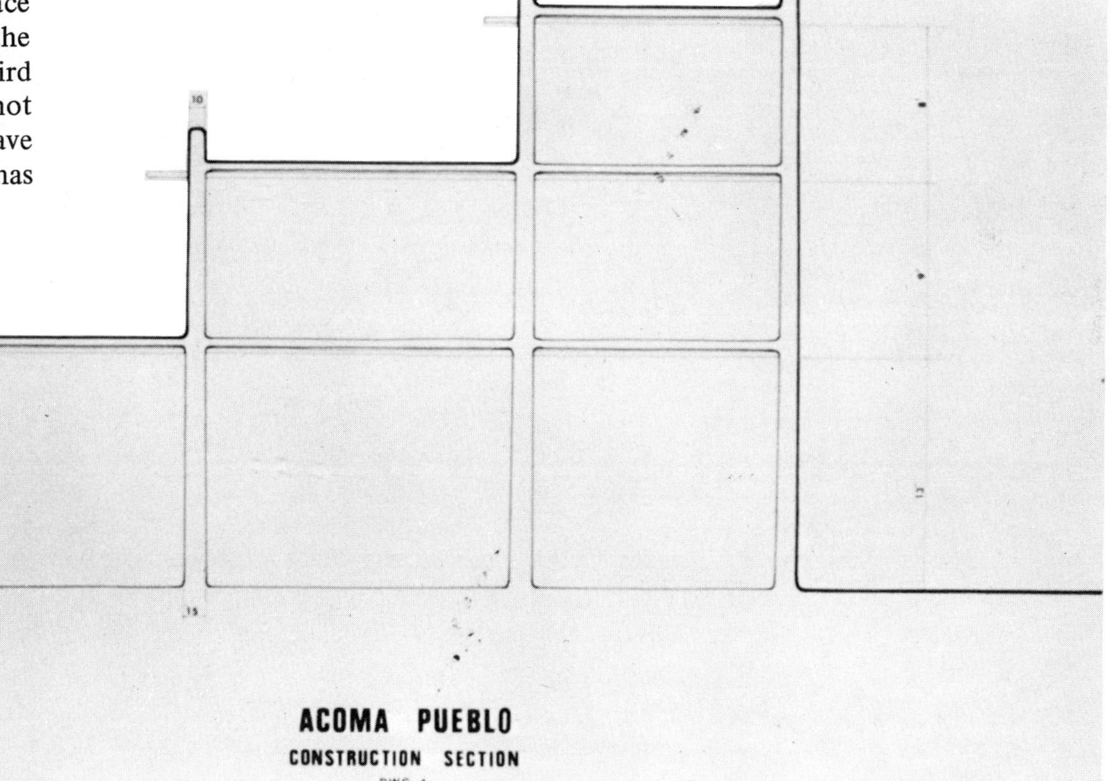

5–13

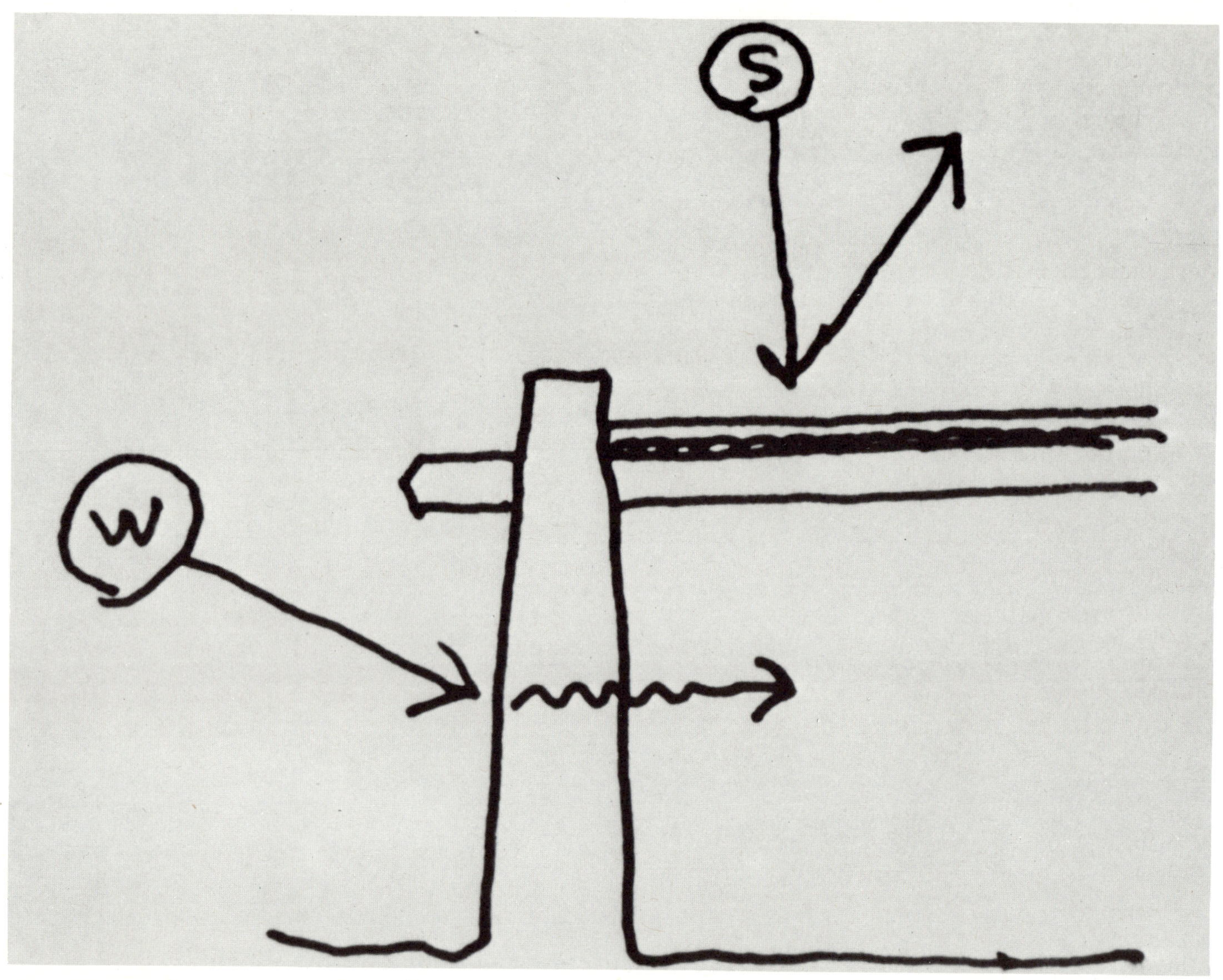

5–14

The low winter rays of the sun strike the sides of the buildings (Figure 5-14) and thus there are vertical, masonry walls with a high heat transmission coefficient and a high heat storage capacity. These walls, in fact, both store and transmit the winter sun, irradiating warmth to the interior of the building and thus to the people within. The seasonal migration of the sun, however, takes it on a different path in summer; then it strikes the buildings from high in the sky, on primarily horizontal surfaces. These are essentially insulating surfaces, made of timber, reeds, grasses and clay. Thus the design keeps the heat out in the summer and transmits it in winter. (Figure 5–15)

In order to make that building design work on a community scale, one has to have access to the sun. And the pueblo did ensure access to the sun. We looked and were unable to find a significant example of shadowing by one of these buildings upon another, in spite of the fact that this pueblo had been continuously lived in and modified over a period of a thousand years. While we cannot absolutely prove it, it certainly seems evident that a design ethic is at work here, and a value system based on a relationship to the heat, light, and movement of the sun.

In about 1520, the Spanish attacked the pueblo from the south. They burned a good portion of the settlement, so that the southern portion of Acoma does not conform to the same plan as the part I am describing. But I wonder if the Spanish, who came preaching Christianity and "do unto others," and who burned out this village, had any idea of what these people were about. In fact, they did protect their neighbor's rights; each Acoman ensured the other a right to the sun.

5–15

Our modern subdivision is not nearly as neighborly in this regard. It is frequently based on the Jeffersonian grid in plan. And orientation is to the street, not the sun. The result is that some design options are not explored and that the people who live there also have options denied.

If a property has access to the sun, it offers the potential for development that is responsive to the sun. The site will demonstrate certain performance characteristics in response to the daily and seasonal migration of the sun that the designer may predict. The designer may therefore choose to provide sun and shade as people and activities need or desire it, rather than force them to adapt arbitrarily. Without solar access, these choices are denied.

Today we shadow our own land and our neighbor's homes and land in ways that interfere with land use. The nature and extent of interference varies with orientation and with the time of day and year.

5–16

A long north–south block accentuates daily variations of shadow impact and of land use. Winter morning shadows are in the front yard of the western row and in the backyard of the eastern row. Afternoon shadows are in the backyard of the western row, and in the front yard of those houses along the eastern edge of the block.

Figures 5–17 through 5–23 (part "a" of each) show the patterns of shading on such a block in Los Angeles, from 9 am in the winter until 3 pm. The result can be a kind of daily migration of outdoor and indoor activities from east to west, following the sunlight. The cycle would repeat itself each morning to the front, and each afternoon to the back, or perhaps the reverse, depending on how the house faces the street. Decisions about where to locate a sandbox or a child's swing, or where to get a suntan, would be influenced by such factors.

Inside the house, the daily tempo of sunlight and shade would be reflected as first one end, then

Winter, long north-south block, 9 AM. 5–17a

Winter, long east-west block, 9 AM. 5–17b

the other end, of the house received its share of morning or afternoon sunlight. These solar conditions may influence important decisions about everyday living patterns. In this orientation, summer shadows behave similarly to winter shadows. But in an east–west block (the "b" parts of the figures), the pattern is not only different from that on a north–south block, but there is an important difference in the summer and winter shadow patterns.

Residents on a block that runs long in the east-west direction have a different set of conditions to which to adjust their activities. On this block, the winter shadows always stay to the north of the house. In the morning, they are cast to the northwest; in the afternoon, they are cast to the northeast. Instead of alternating between the front and back of the house on a daily tempo, the shadows will stay in the front yard of the northern row and the back yard of the southern row all day and all winter long.

The east–west development is subject to a seasonal tempo of sun and shade, whereas the north–south development demonstrates a daily tempo. Summer shadows on an east–west development, unlike winter shadows, are cast to the south of the house. Morning shadows are cast to the southwest and afternoon shadows to the southeast. Only at mid-day do the shadows fall slightly to the north. The result of this shift in the direction of shadows can be a seasonal migration of activities from north to south and back again.

Winter, long north-south block, 10 AM. 5–18a

Winter, long east-west block 10 AM. 5–18b

The difference of tempo based on orientation raises many questions that fall under the general heading of life quality and that affect land use, in this example, of front, back and side yards. Orientation influences the rhythms of light and heat and therefore the way predetermined arrangements are going to perform. To the extent that I can plan or design those rhythms, I can conceive of rhythm as a strategy for design.

We said, in our recent work, that there is a lot of uncertainty in this situation, given present land use models, and we looked for ways to deal with that. Could one reduce this uncertainty by limiting the impact of one of these constructions upon another construction, and upon the site of another construction? This led us to the whole issue of sun rights, solar access and, ultimately, quite straightforward solar zoning.

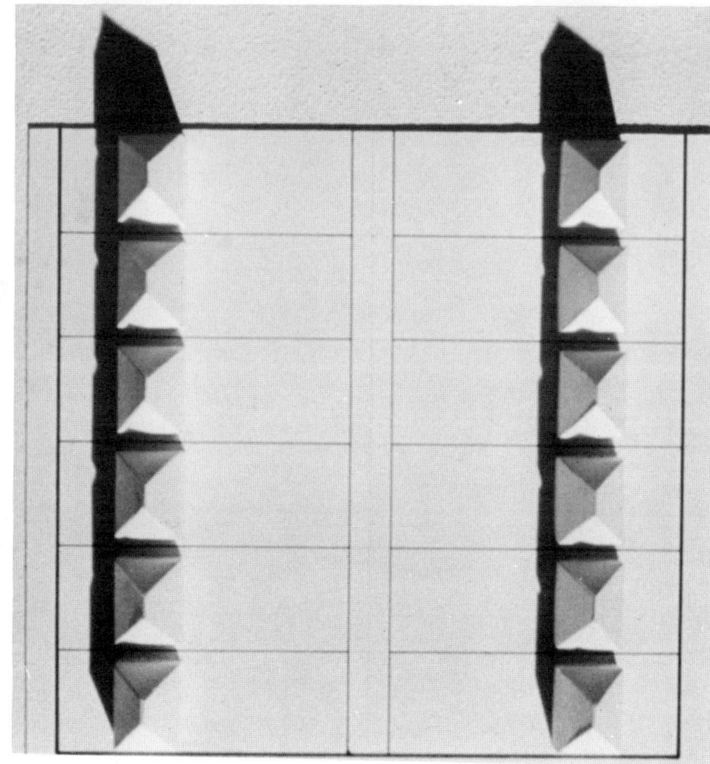

Winter, long east-west block, 11 AM. 5–19a

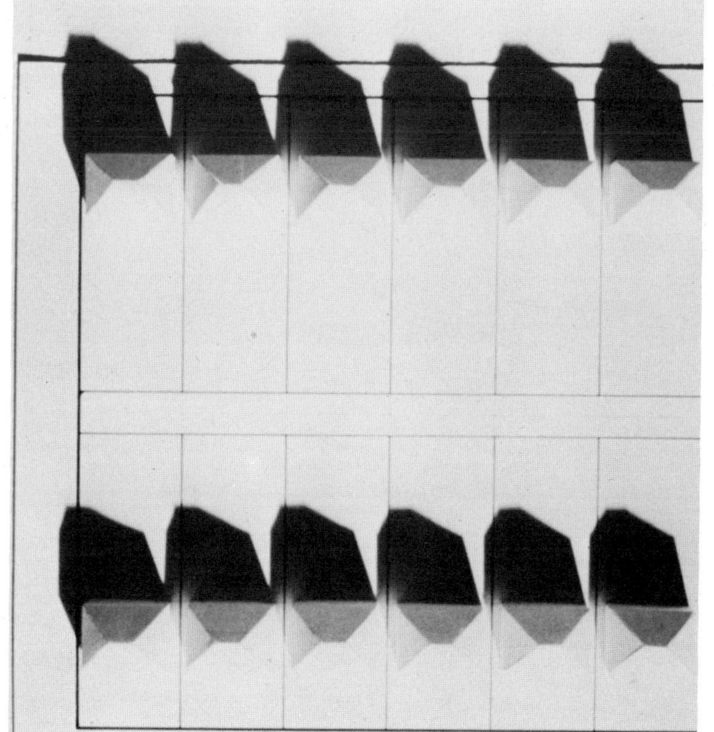

Winter, long north-south block, 11 AM. 5–19b

Winter, long north-south block, noon. 5–20a

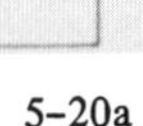

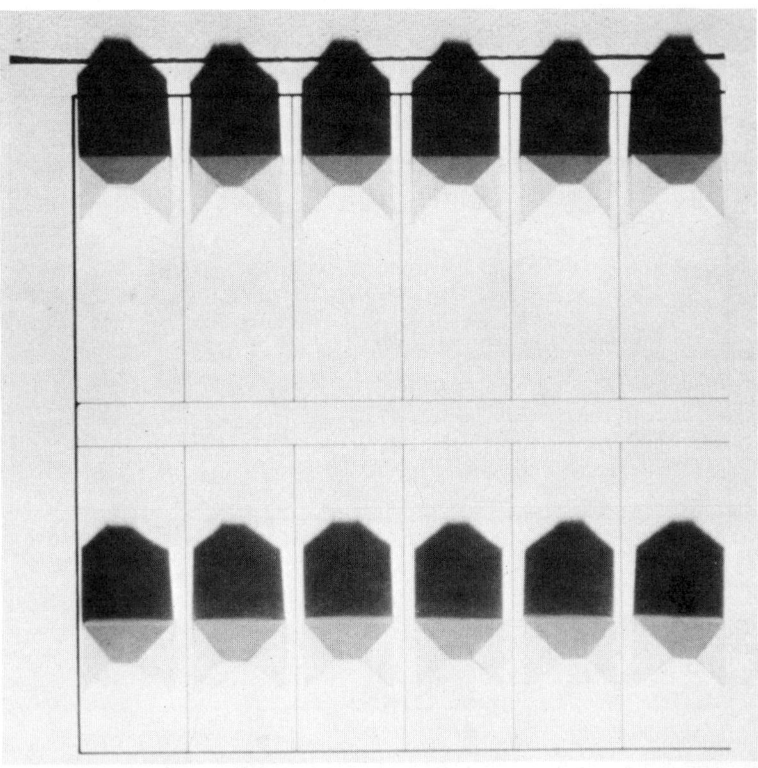

Winter, long east-west block, noon. 5–20b

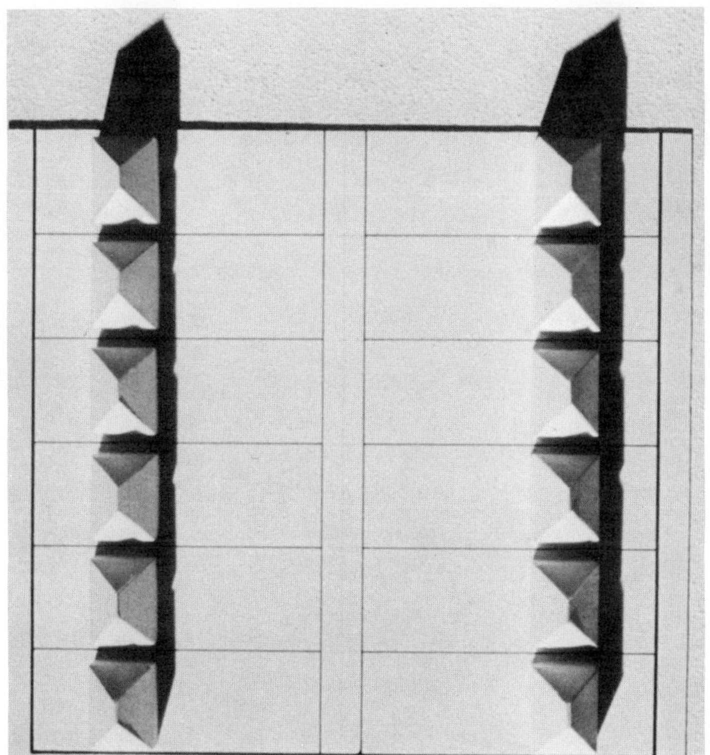

Winter, long north-south block, 1 PM. 5–21a

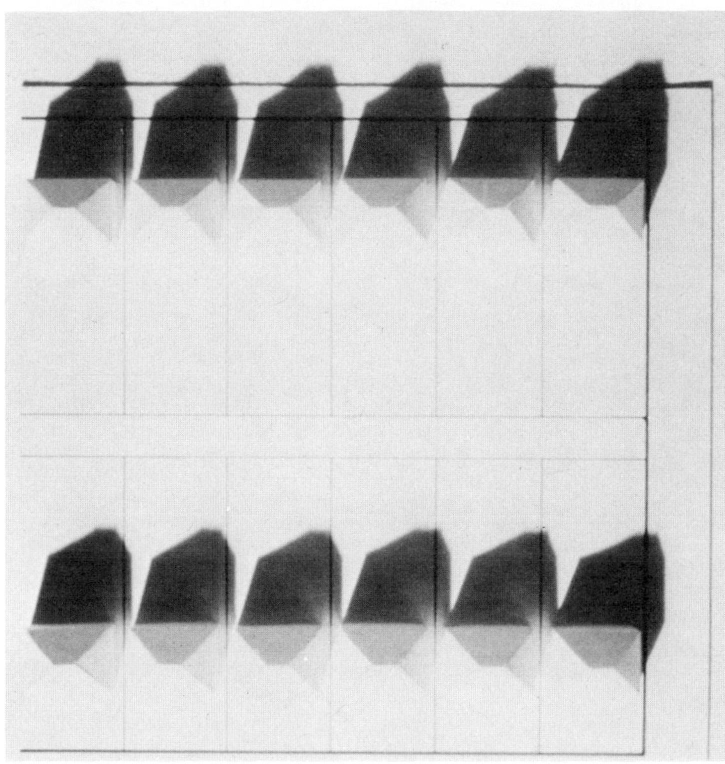

Winter, long east-west block, 1 PM. 5–21b

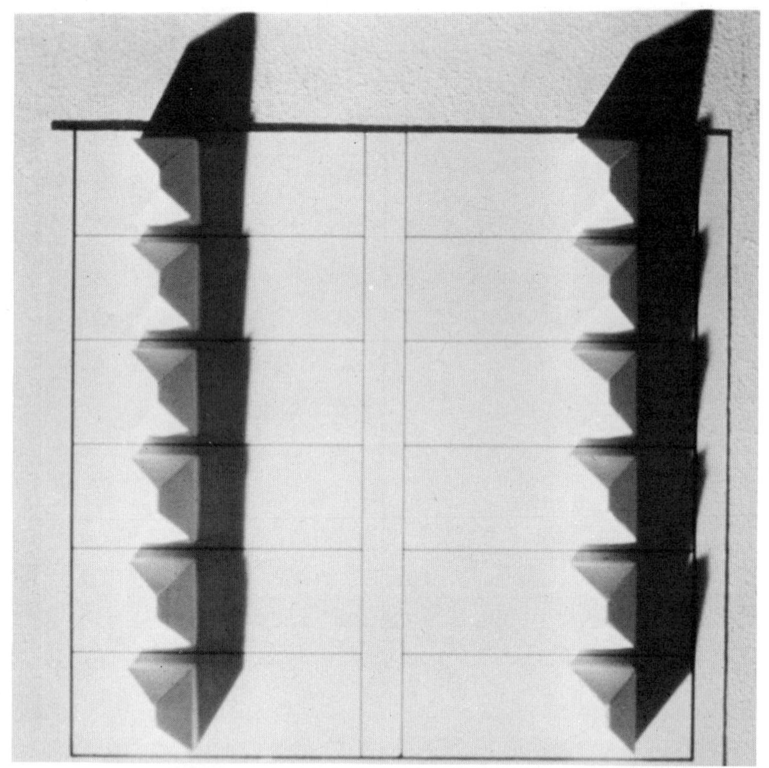
Winter, long north-south block, 2 PM. 5–22a

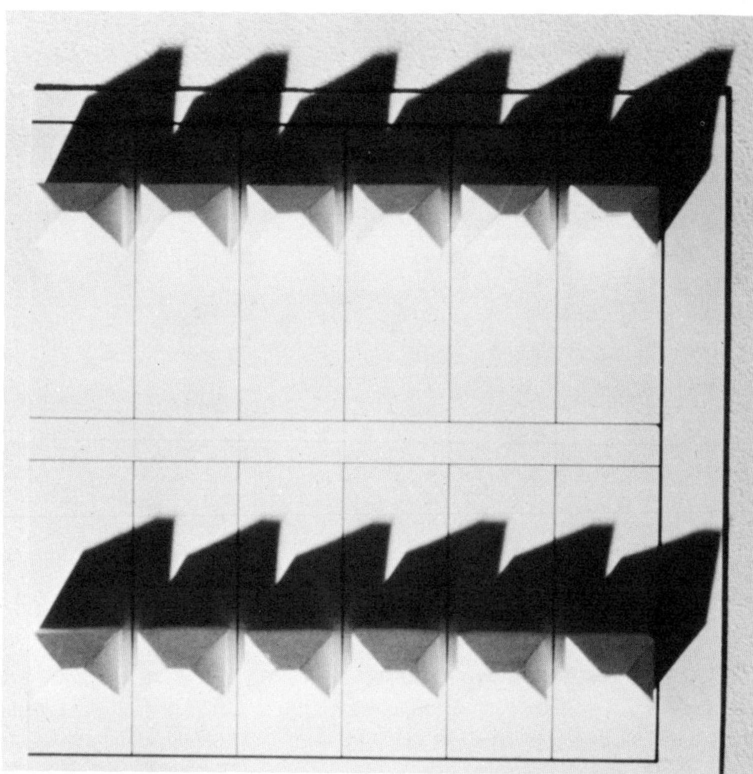
Winter, long east-west block, 2 PM. 5–22b

To provide an image of solar zoning, I'm now going to replace those small developer houses with envelopes of volume, defined by ridges. The ridges are as large and high as they can be without shading a neighbor's property during specified periods of time.

Figures 5–24 and 5–25 illustrate a site repre-

Winter, long north-south block, 3 PM. 5–23a

Winter, long east-west block, 3 PM. 5–23b

senting a whole city block. One half of the site is subdivided into smaller lots with 50-foot frontages and the other half is left as a single parcel 300 feet long. In this case, the ridges indicate the maximum volume of space that could be developed on these sites without shadowing a neighbor between 9 am and 3 pm in the winter and between 7 am and 5 pm in the summer (when the sun is higher in the sky and therefore casts shorter shadows). Figure 5-24a shows the envelope at noon; Figure 5-25a shows the same site at 3 pm (winter), when the shadow reaches precisely the edge of the lot.

Maximum development potential for any solar envelope is described when the shadows reach the perimeter of the land parcel at cut-off times. If the shadows at cut-off times did not reach the perimeter of the land parcel, we would not have defined all the available volume. If the shadows went outside the perimeter, we would have defined more volume than we had a right to define.

Winter noon. 5-24a

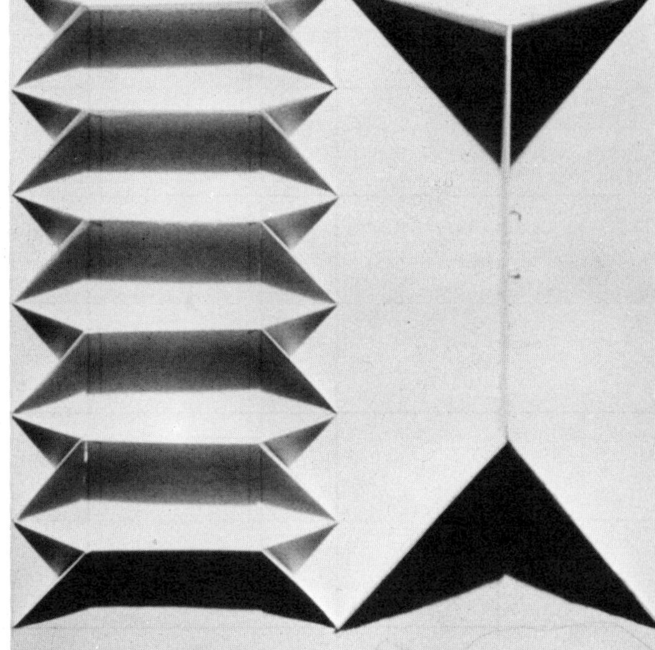

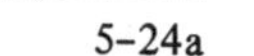

Winter noon. 5-24b

Figures 5–24b and 5–25b demonstrate the same phenomena for properties in the opposite orientation.

The solar envelope is a logical and rational construct of time and space defined by the movement of the sun. The daily east-west migrations of the sun describe the east and west boundaries of the envelope. The seasonal migration of the sun, from low in the winter sky to more northward in summer, describes the north and south boundaries of the envelope. The envelope is dependent on the time of day and season and it is dependent on the shape and orientation of the grid. Or, to turn it around, it is as much a function of time as it is of space.

The orientation of the land parcel has particular significance for the shape of the solar envelope. A site that runs long in the north–south direction produces an envelope with a long ridge in the north-south direction that lies symmetrically between the

3 PM winter. 5–25a

3 PM winter. 5–25b

two edges of the land parcel. A site that runs long in the east–west direction produces an envelope with a long ridge in the east–west direction, but the ridge is asymmetrically located on the site, closer to the south than to the north boundary. In addition, the envelope that runs long in the north–south direction would tend to expose major building surfaces to the east and west and therefore to a daily rythm. The envelope with an east–west ridge suggests major building surfaces facing north and south and therefore exposure to a seasonal rhythm. If the Piutes were right, if the Longhouse builders

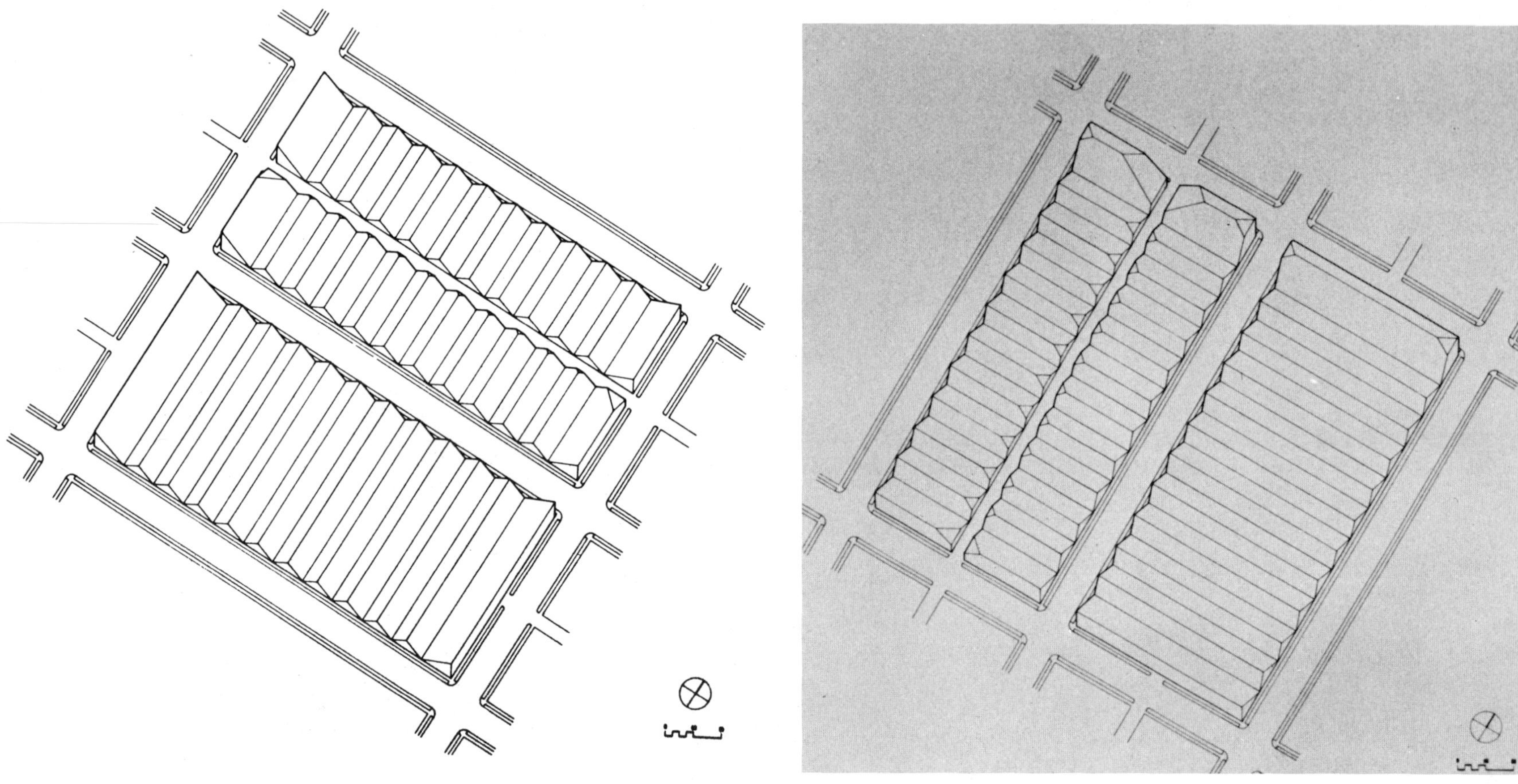

50-foot frontage. 5–26a

50-foot frontage. 5–26b

were right and if the Acomans were right, there is a certain advantage to the north-south building orientation because it provides a large south exposure. In addition, the envelope with an east-west ridge provides more developable volume than the envelope with a north-south ridge.

The solar envelope is intended to limit development that would deprive a neighbor of access to the sun, but how the envelope is applied and at what scale is a matter of design and of policy. Land assemblage is one way to increase the developable volume provided by a solar envelope; the larger the

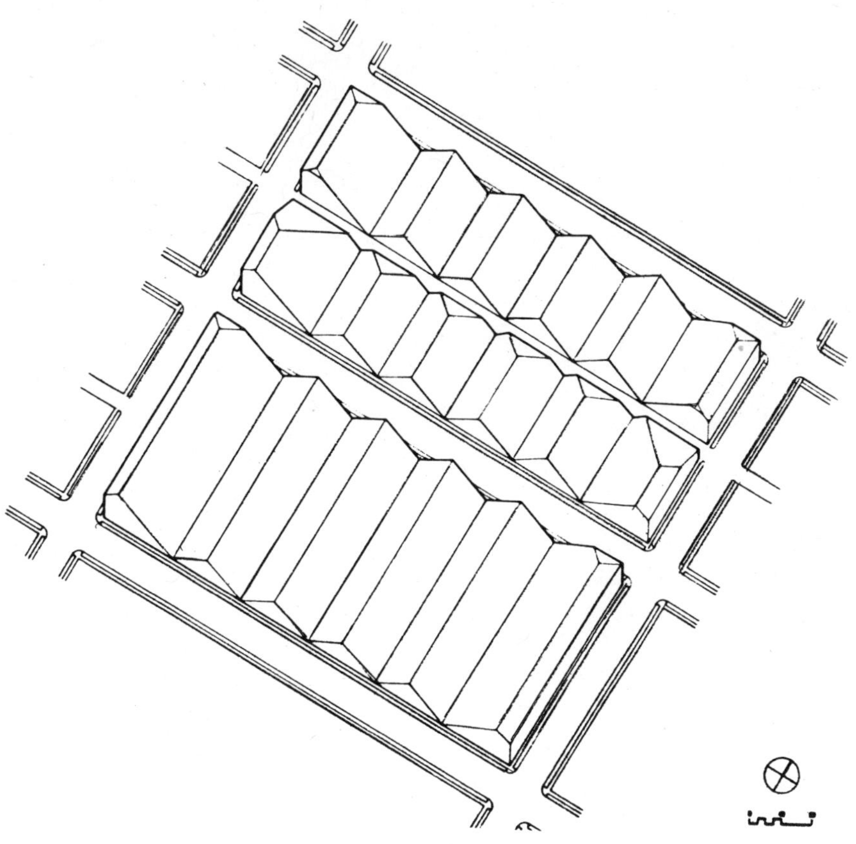

100-foot frontage. 5-27a

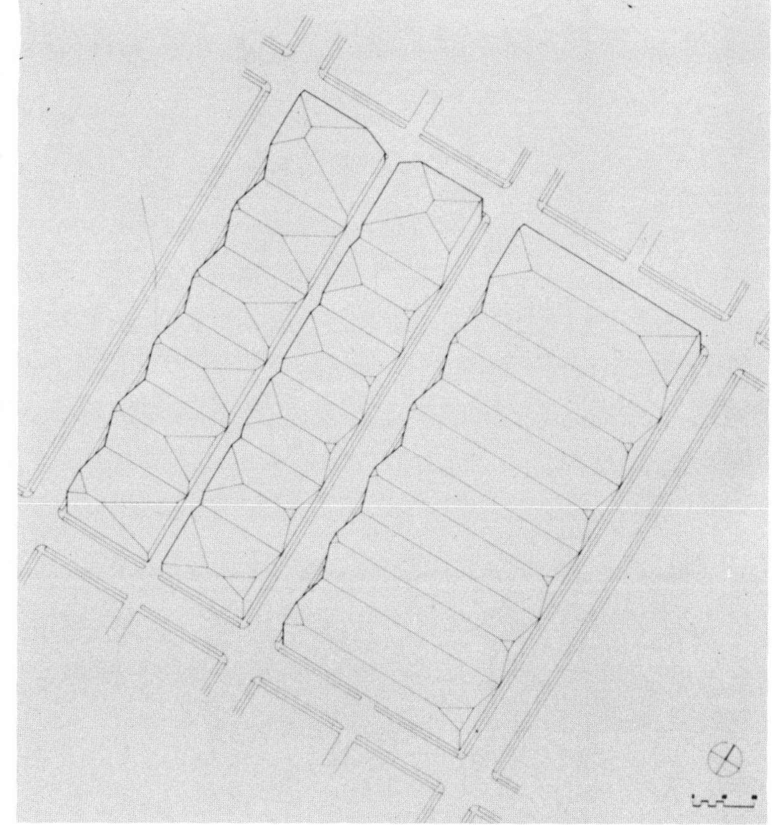

100-foot frontage. 5-27b

land, the greater the volume of development. But, from the point of view of solar access, land assemblage is not always good.

If we assemble the 50-foot frontage lots into a larger land parcel, the orientation of the resulting property changes, and so does the orientation of the solar envelope. If the 50-foot lots run long in the east/west direction, their individual solar envelopes will have broad north and south exposures. But if the lots are assembled, the aggregate property will run long in the north-south direction, and its solar envelope will have broad east and west ex-

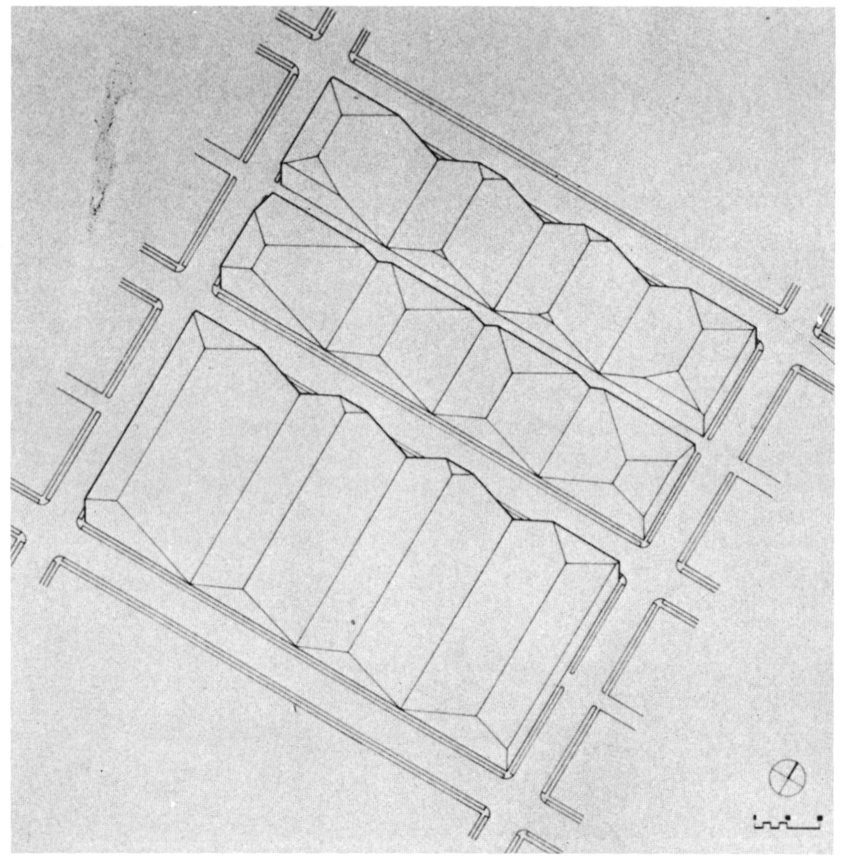

150-foot frontage. 5–28a

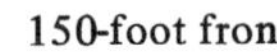

150-foot frontage. 5–28b

posures. As a result of land assemblage, the solar envelopes change from a seasonal exposure to a daily exposure, with implications for the development within.

If the same principles of land assemblage are applied to 50-foot lots that run long in the north–south direction, the orientation of the aggregate envelope will improve. The individual envelopes in this case have broad east–west faces, or a daily tempo; the solar envelope for the assembled parcel of land gains broad north–south faces, or a seasonal tempo.

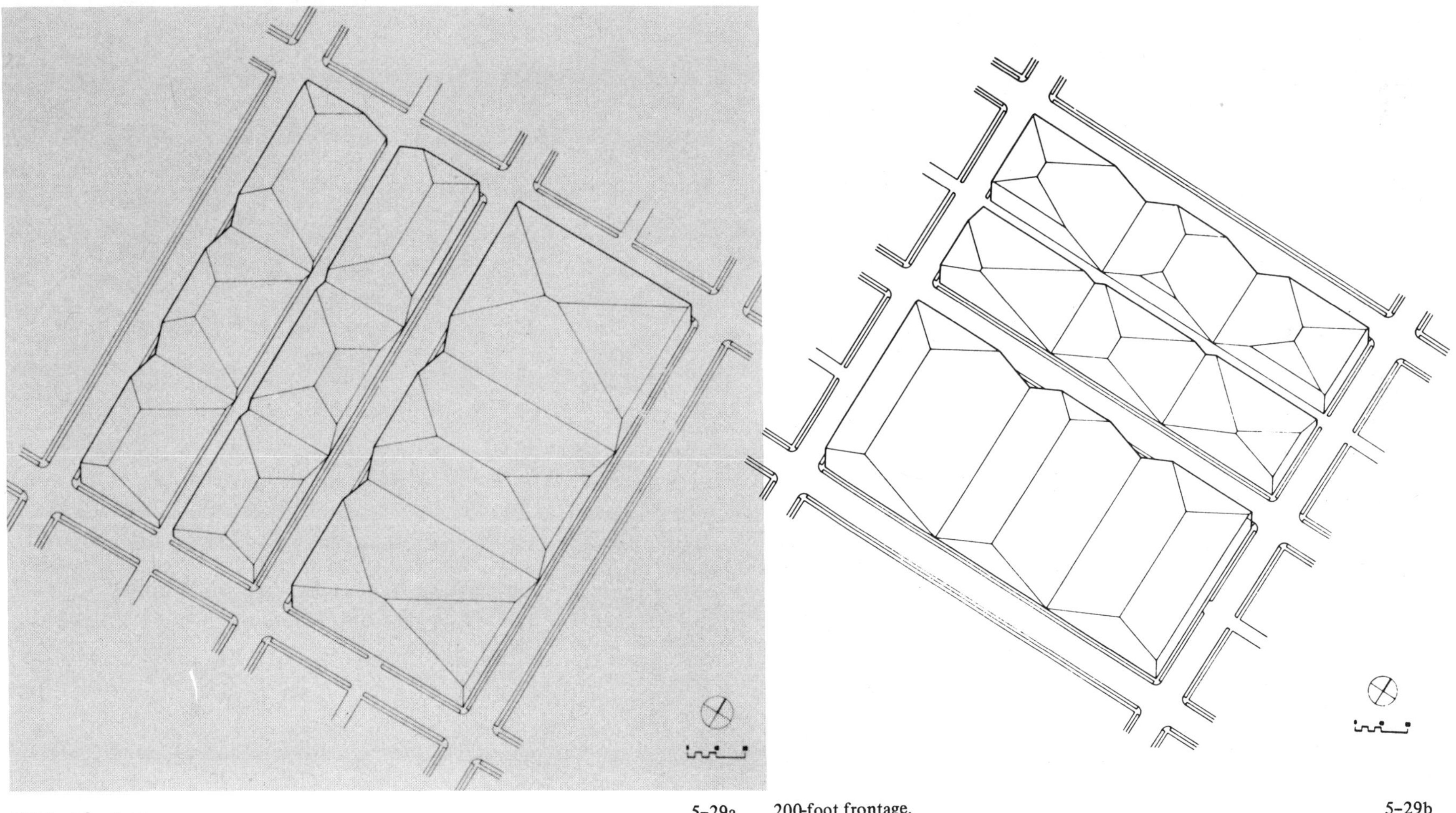

200-foot frontage. 5–29a

200-foot frontage. 5–29b

Obviously, we are here dealing with a situation where the changes in orientation produce the potential for solar development with different beats. The pulses of one orientation are not the same as the other. Any design effort that takes place within the constraints of a solar envelope must recognize rhythm as a design issue.

We generated solar envelopes for all of the orientations just described. In Figures 5-26 through 5-31, the ridges of the envelope have been filled out and the resulting form resembles a tent. The tents describe the volumes defined by the ridges; development within these tents, or envelopes, will not shadow a neighbor during critical times. The time constraints for these envelopes have not changed from the previously described examples, but the spatial constraints have changed. These envelopes are allowed to shadow across a street to the edge of neighboring property on one side, and across an alley to neighboring property on the other side.

Figures 5-26 through 5-31 include blocks of north-south and east-west orientations, subdivided into lots with frontages of 50 feet, 100 feet, 150 feet, 200 feet, 300 feet and 600 feet. There is considerable variety. Note that the corners are in all cases different, with distinctive urban design implications. The corners are unique because they are allowed to shadow across the road in two directions. Corners, therefore, become special events and exhibit special design properties.

The formal implications of changing orientation and scale are dramatic. The crystalline forms emerge as unique. Where there is a series of only three envelopes on a land assemblage, with one inside envelope and two outside envelopes, the variety is especially rich, and represents maximum diversity.

Changes in orientation not only create formal interest, but have importance for energy conversion. As we assemble land parcels from 50-foot frontages, to 100 feet, to 150 feet, the polarity of the envelope ridges starts to change. Envelopes with north-south faces start to give way to envelopes with east-west faces; the reverse happens in the opposite orientation.

To gain south exposures in an envelope with east and west faces, it would be necessary to cut up the block like a piece of salami and, in the cutting process, volume would be lost. An envelope with north and south faces, on the other hand, can achieve full potential for energy conversion without losing volume.

It is possible, therefore, to draw the conclusion that blocks running long in the east-west direction lend themselves to development in small lots. They then start with solar envelopes having broad north-south exposures, which are able to provide 30 to 40 percent more volume than an envelope for a similarly-sized lot running long in the north-south direction. But as the east-westland parcels are assembled, the ridge of the envelope turns and the south-face advantage is lost.

Properties running long in the north-south direction lend themselves to development at large scale. Divided into small lots, they yield solar envelopes with east-west exposures, but as the lots are aggregated into larger development sites, the ridge of the solar envelopes turn to yield broad exposures to the north and south, and thus a more sympathetic orientation for energy conversion.

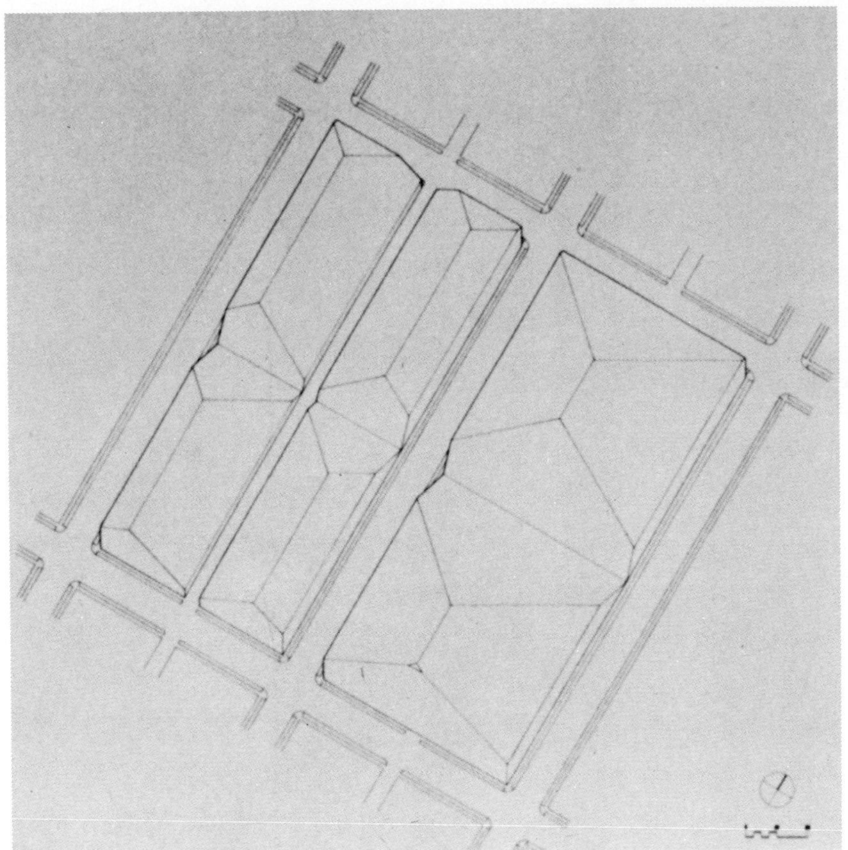

5–30a 300-foot frontage.

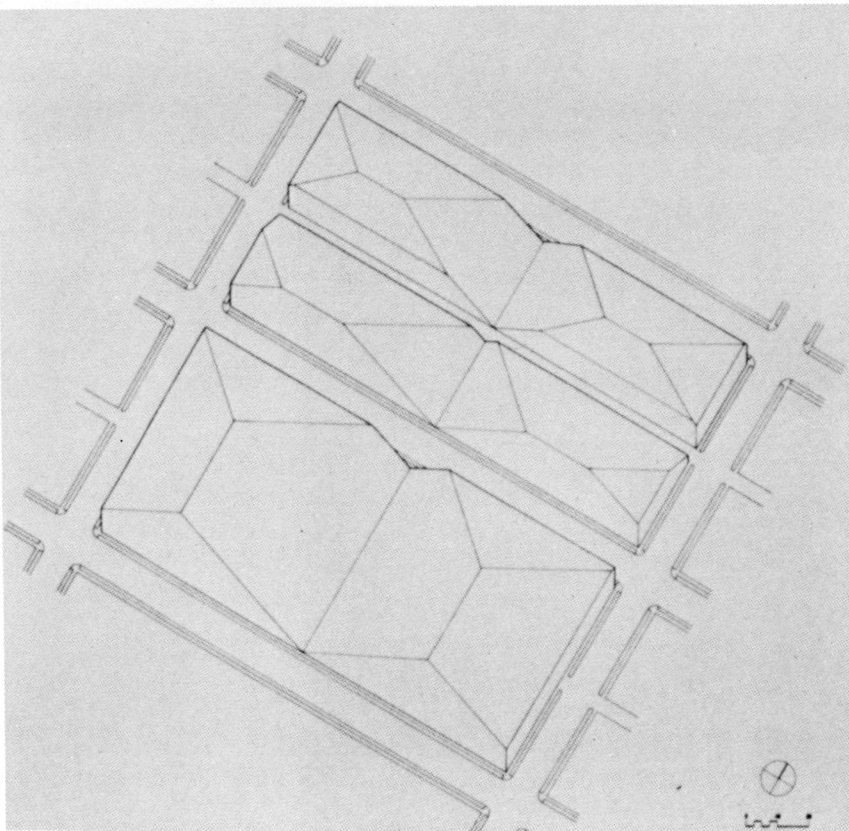

300-foot frontage. 5–30b

If we plot south-facing surfaces of solar envelopes in relation to land assembly for lots running long in the north–south direction versus those lots running long in the east–west direction, the relative advantages of each orientation are clear (Figure 5–32). The initial advantage of one orientation soon gives way to the aggregate advantage of the other.

The importance of orientation becomes even more evident when one tries to design within a

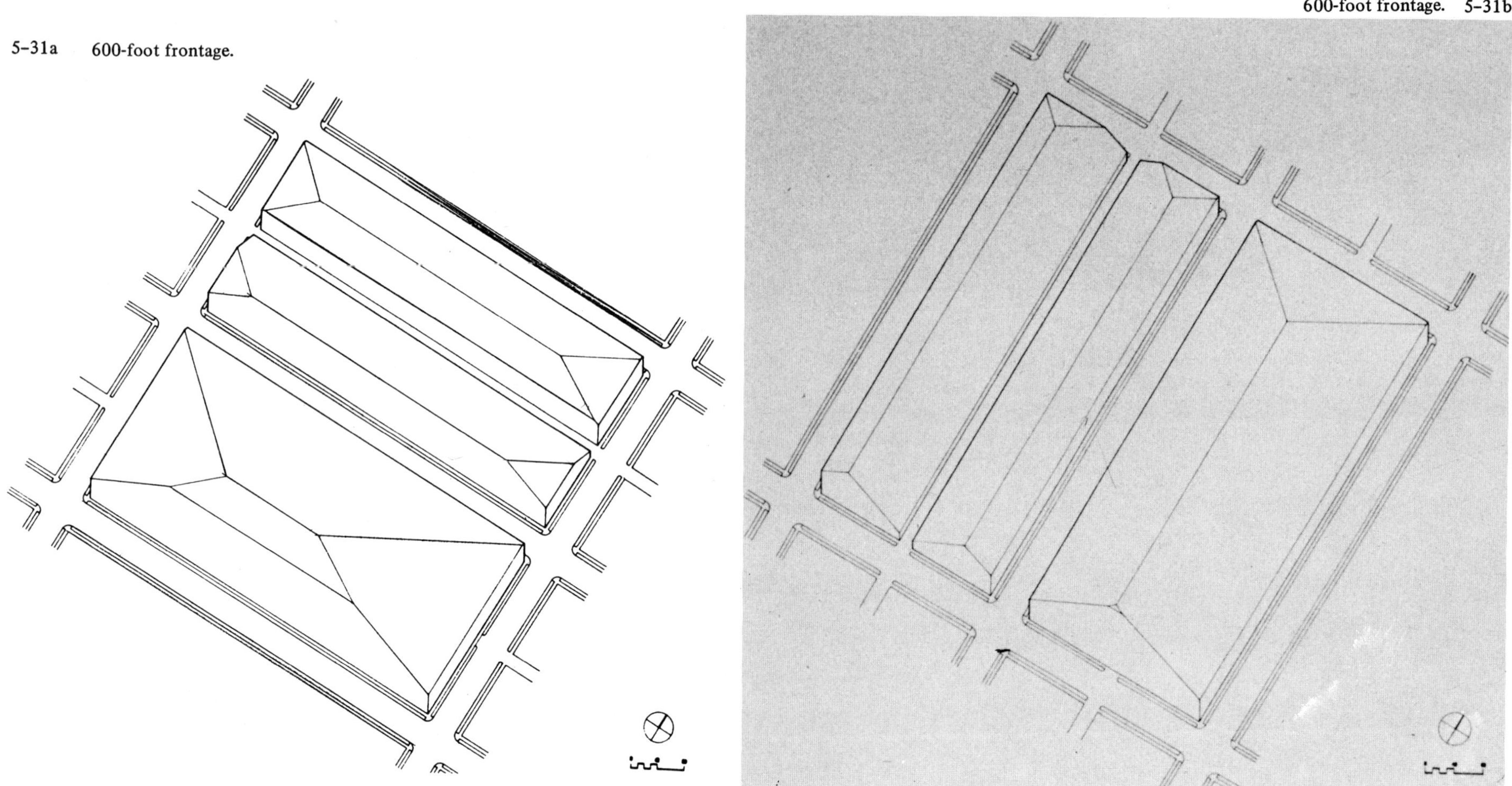

5–31a 600-foot frontage.

600-foot frontage. 5–31b

solar envelope for a small site. The lot in Figure 5-33a runs long in the north-south direction, and, in order to get that critical south exposure, I really had to cut it up—cut right into it. In the process, I got a kind of courtyard house. But I used up a lot of land, and I've got a lot of land that is not used well.

On the east-west lot (Figure 5-33b), however, I

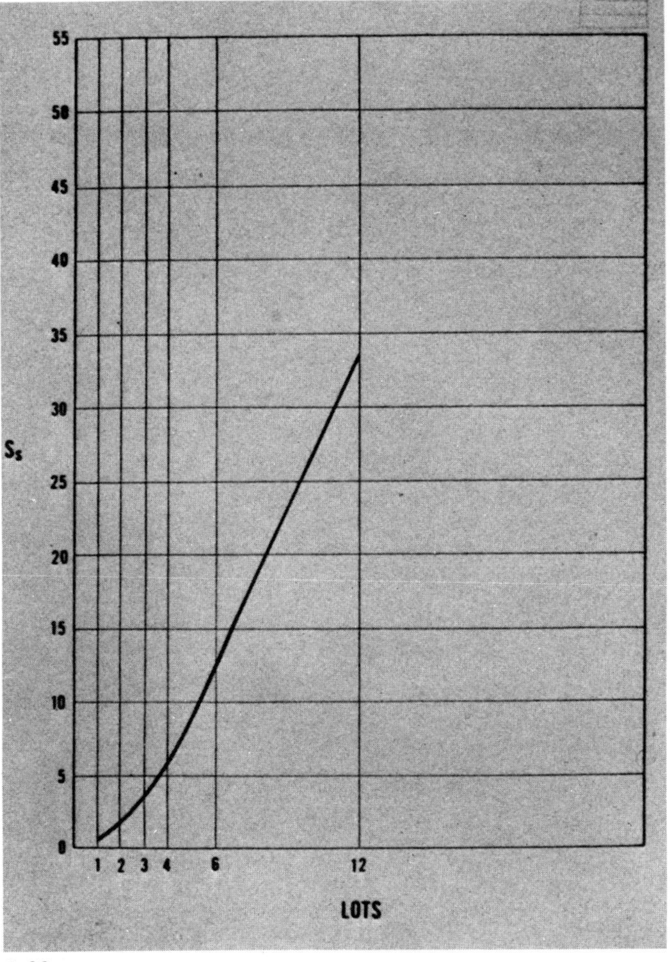

5-32a

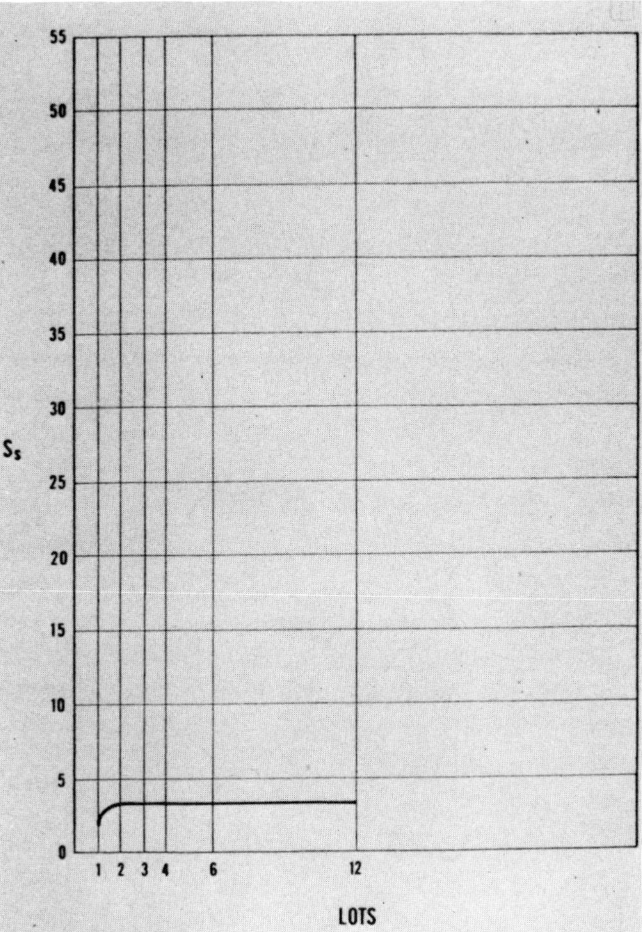

5-32b

have a solar envelope with broad north and south exposures. And I have a ridge height that is 30 to 40 percent higher. I have more volume to play with, and a southern exposure, and therefore I can enjoy much more choice as a designer. Here, my house has a large, unified yard and a lot of possibilities I can arrange differently. From a user's point of view, the advantages of this orientation over the other are clear; this orientation provides more choice.

At this point, I'd like to show you some of the dynamics of the solar envelope: I'd like you to see the envelope as the sun sees it. The solar envelope is, in fact, a set of limits that can be quite conventionally drawn and surveyed, but that have an aspect of time.

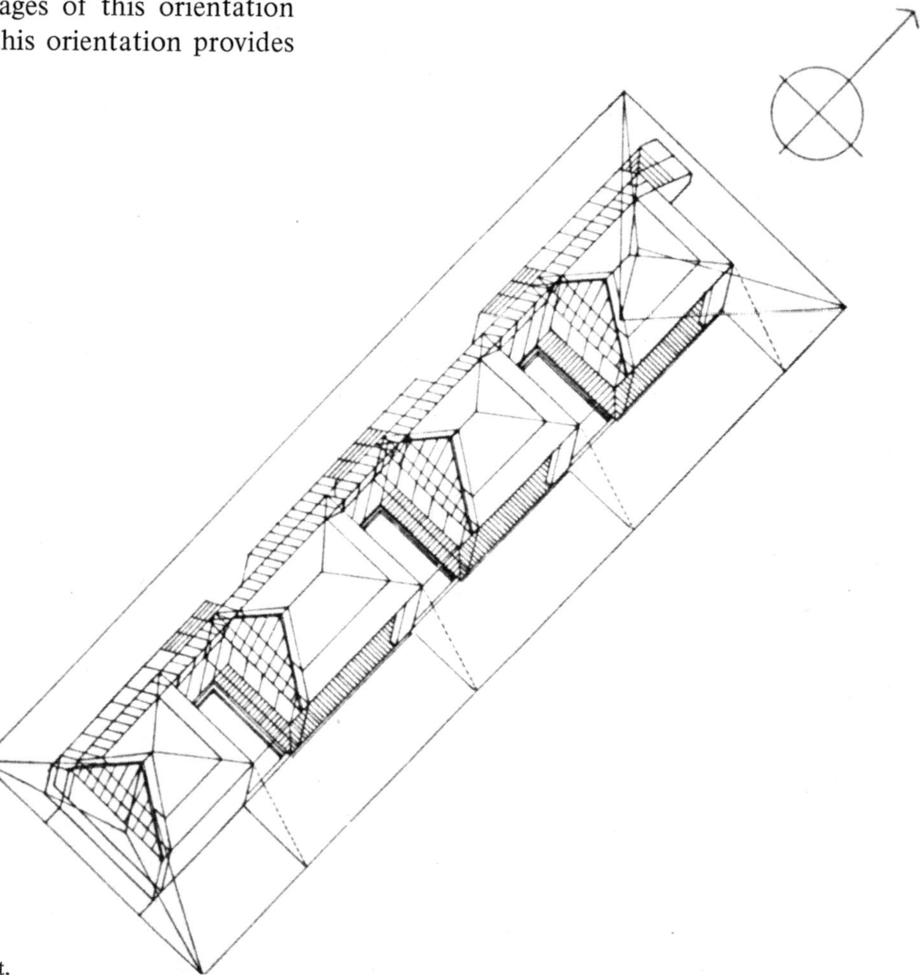

5–33a Long north-south site; house 2500 square feet.

In Figures 5–34 through 5–38, I am seeing the solar envelope as the sun sees it. The land is a 150-foot parcel, facing the street. The series starts with a look at the solar envelope at 9:00 on the morning of winter solstice. I can see the south face of the envelope and the east face. The street face has been cut off by standard zoning limits, and in back is a vertical face that has been cut off by the alley. I cannot see the west face because it is parallel to you, the rays of the sun.

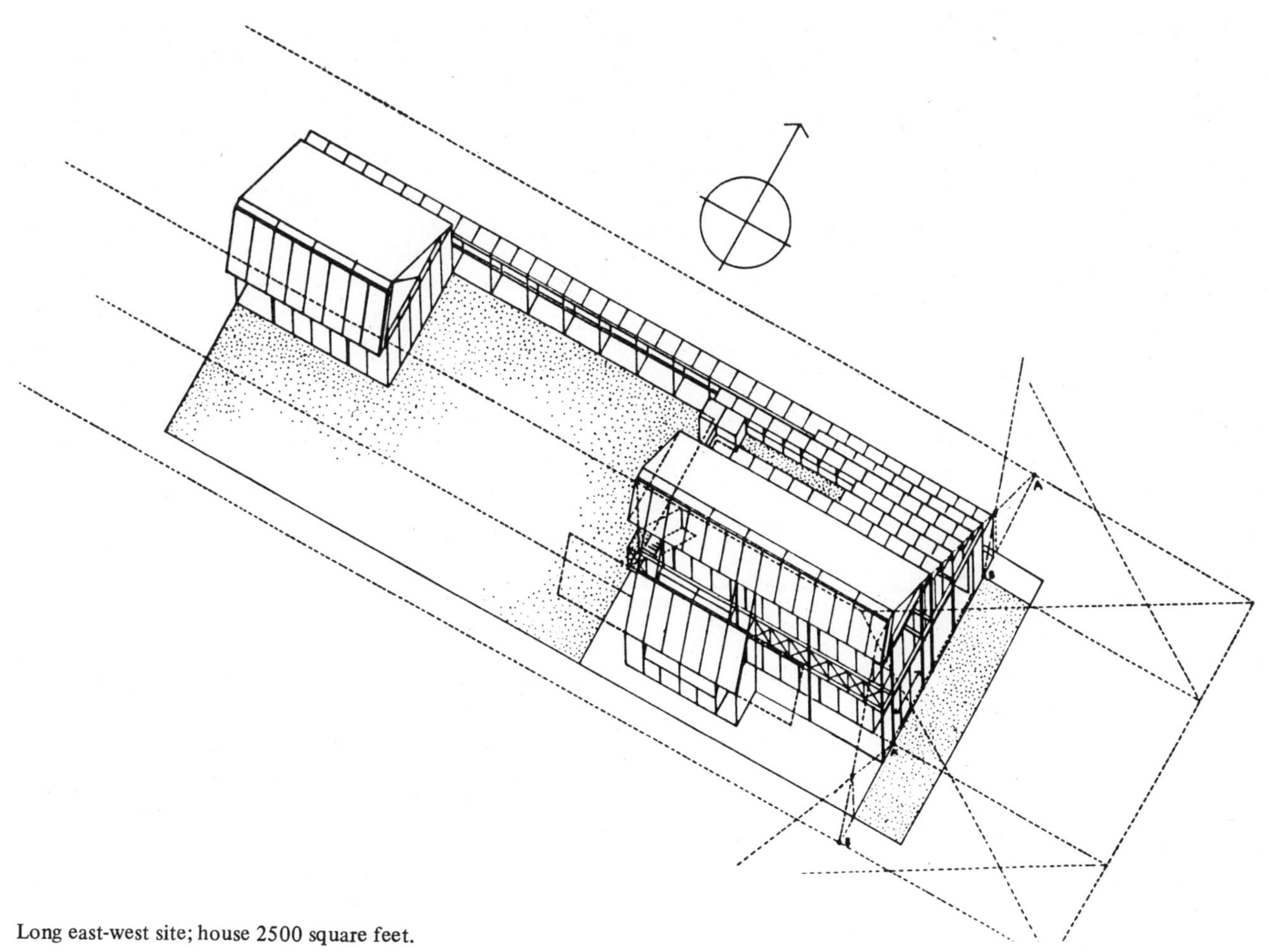

Long east-west site; house 2500 square feet.

5–33b

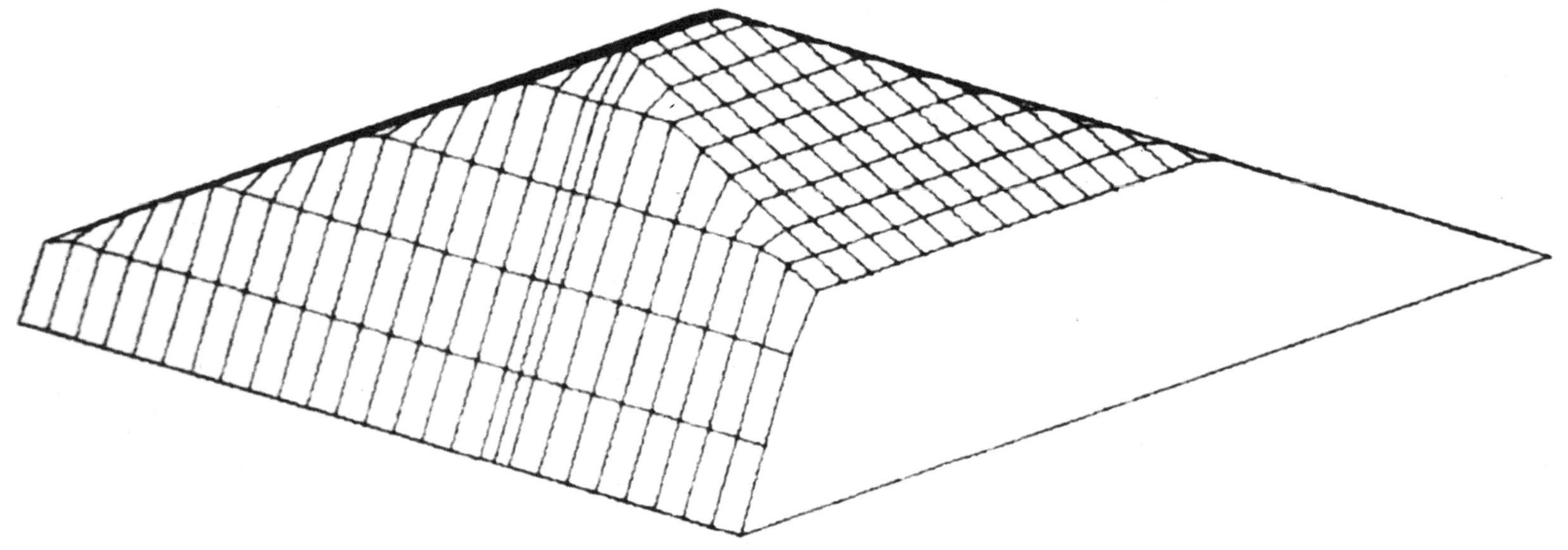

Winter solstice, 9 AM.

5–34

As time passes, the sun shifts in its position and you, as the sun, can see the west and north. As time passes, you move in relation to this form, and at noon you can see the solar envelope dead on, all four faces. Remember here that, as the winter sun, you are looking most directly at the south face of the envelope. By 3:00, you are again parallel to the critical faces. The north is critical in wintertime, as is the west face in the morning and the east face in the afternoon.

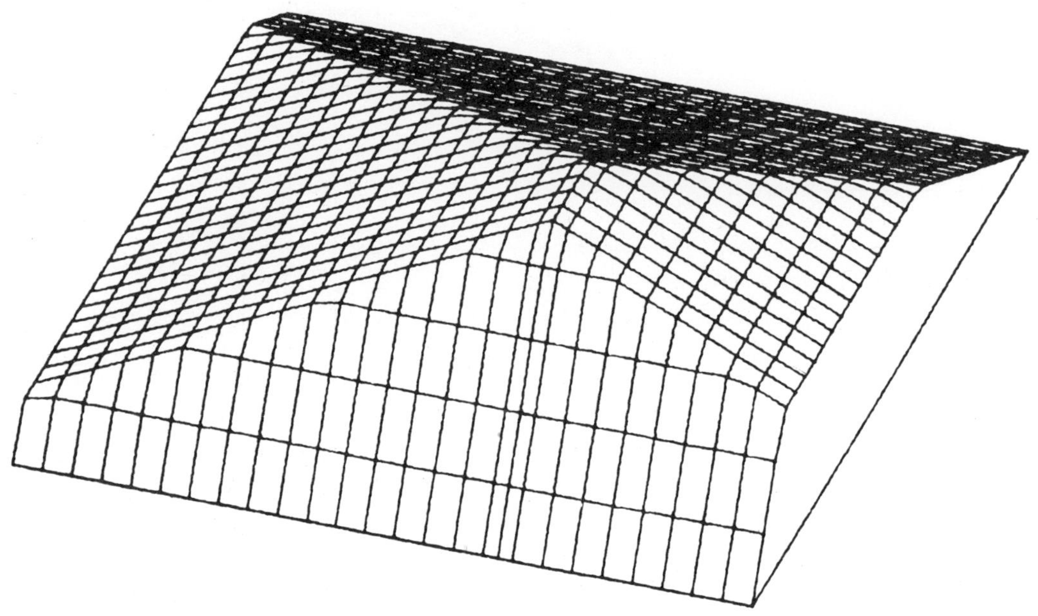

5-35 Winter solstice, 11 AM.

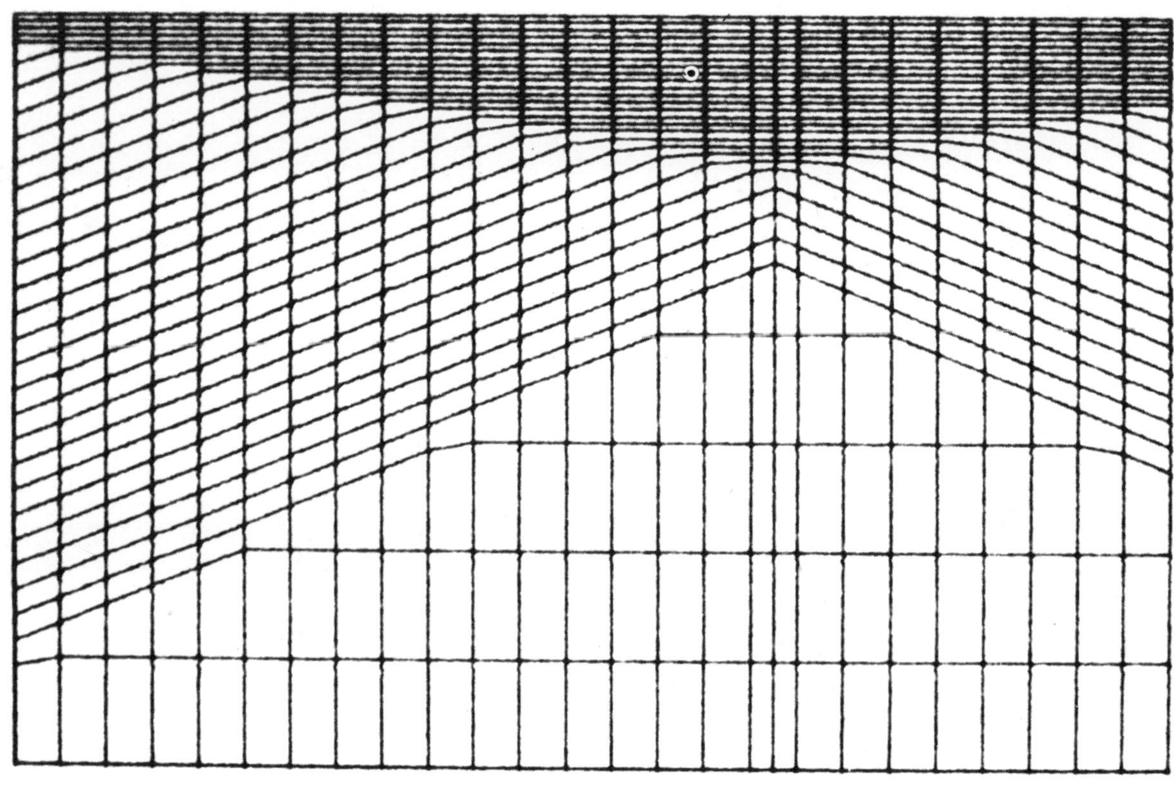

5-36 Winter solstice, noon.

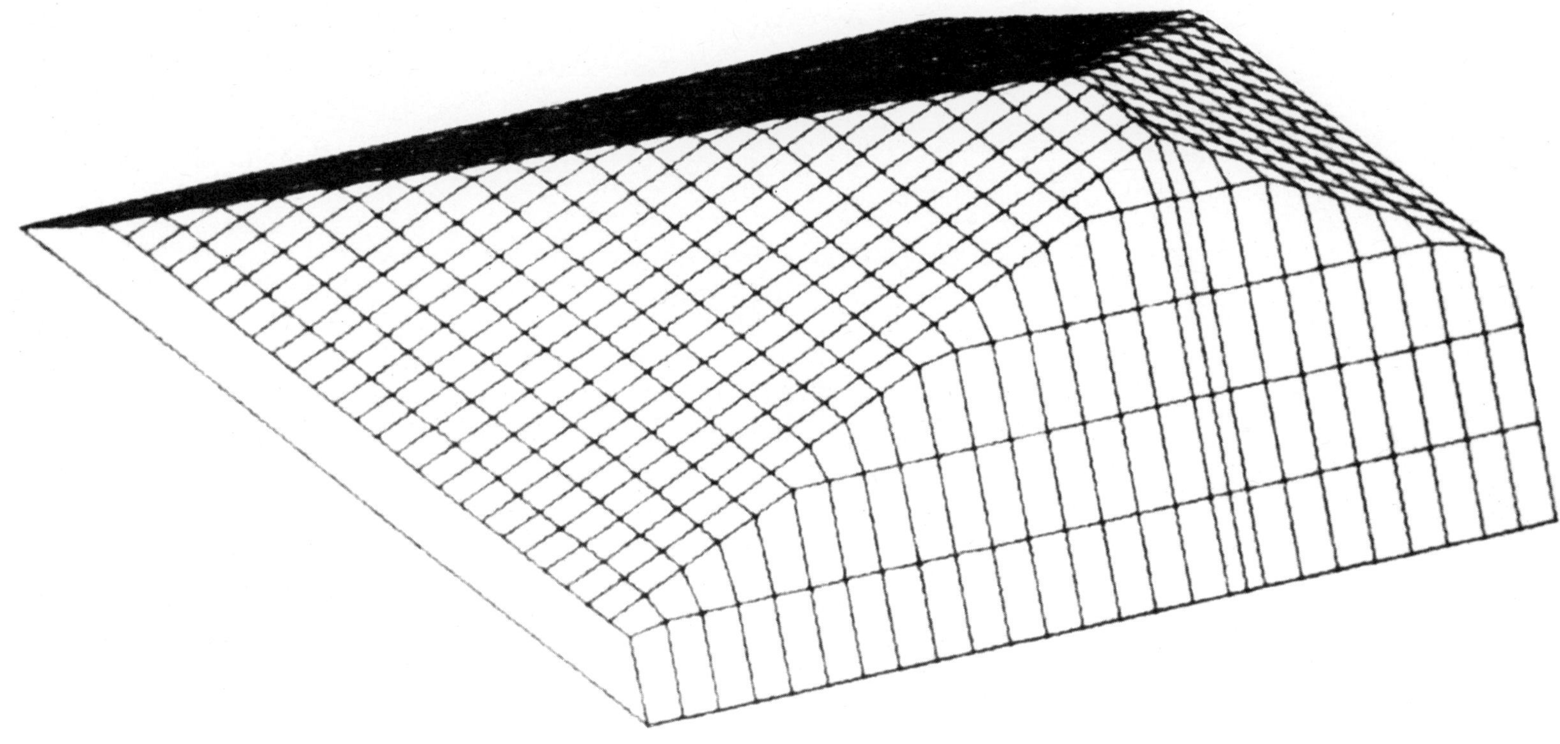

Winter solstice, 2 PM. 5–37

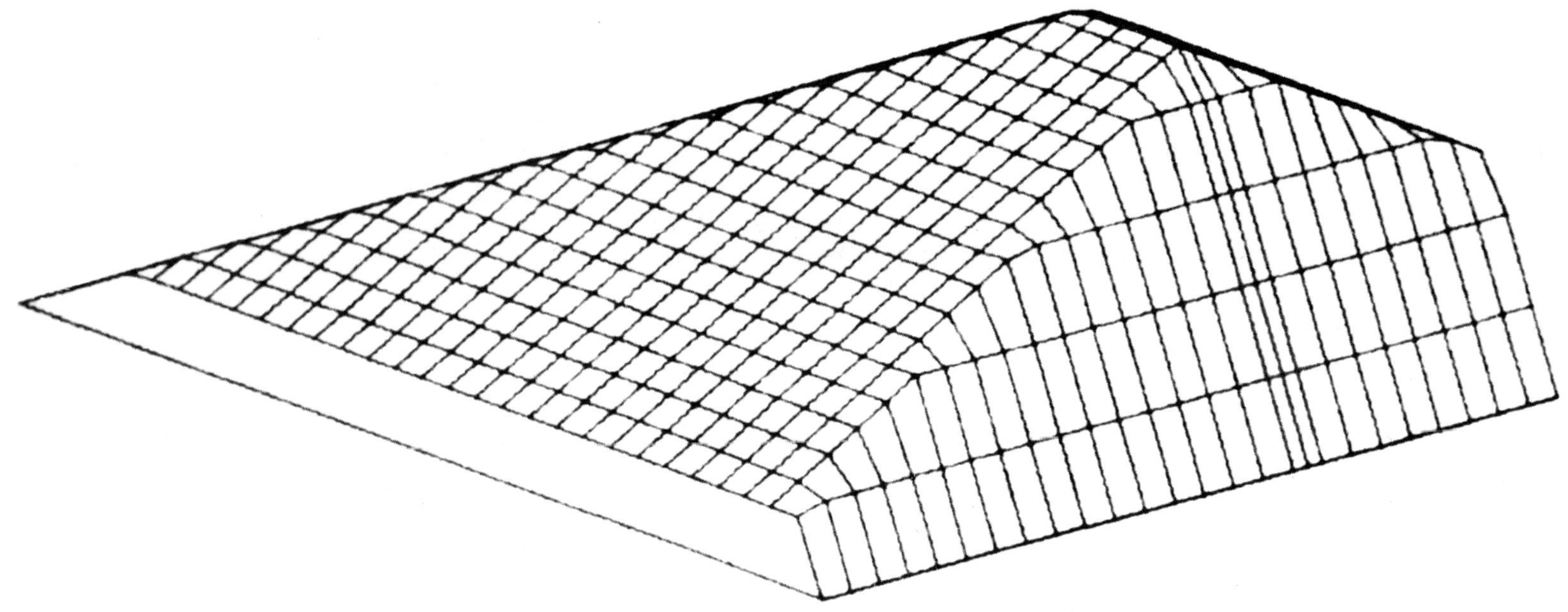

Winter solstice, 3 PM.

5–38

Figures 5-39 through 5-45 show the envelope as the sun would see it at summer solstice. The sun rises to the north of east; the critical faces are therefore south and west. By 8:00, you, as the sun, can see the south face. By 9:00, you can see the south face straight as you move across the east/west axis of the cardinal points. But, at mid-day, you rise high above the south face and so do not see it directly.

From 11:00 until 1:00 in the summer, the sun goes through the same azimuth angular rotation as it does from 9:00 to 3:00 in winter. That is why the south exposure is critical. It receives maximum exposure to the sun in winter, when heat is desirable. It receives minimum exposure to the sun in summer, when heat is not desirable. This is important not only for energy conversion, but ultimately, for life quality. But as the designers became more

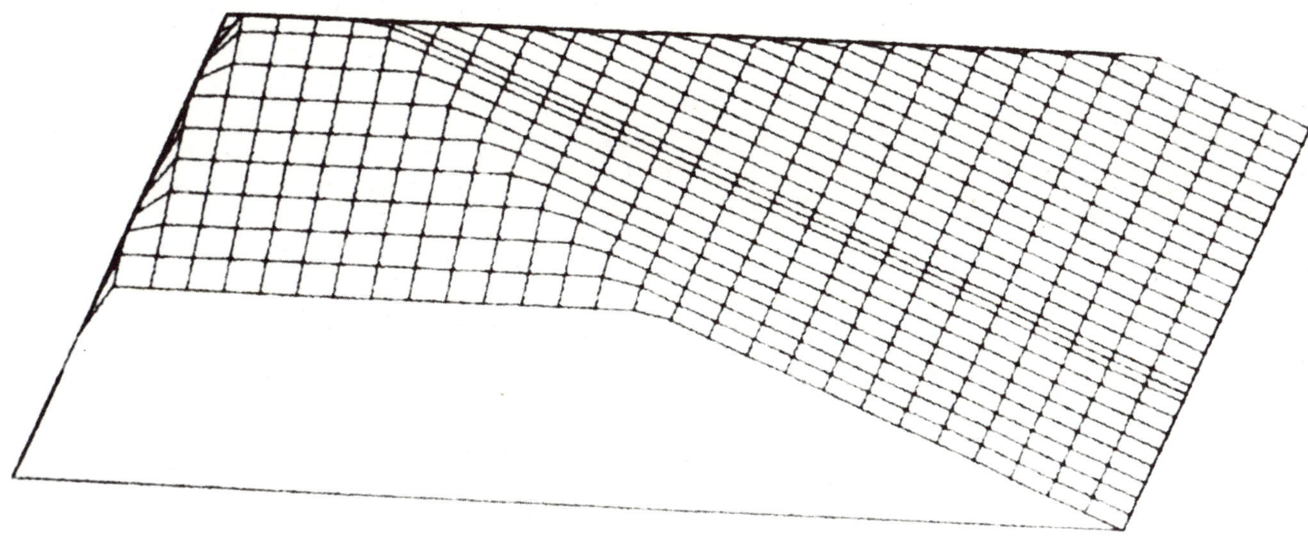

5–39 Summer solstice, 7 A.M.

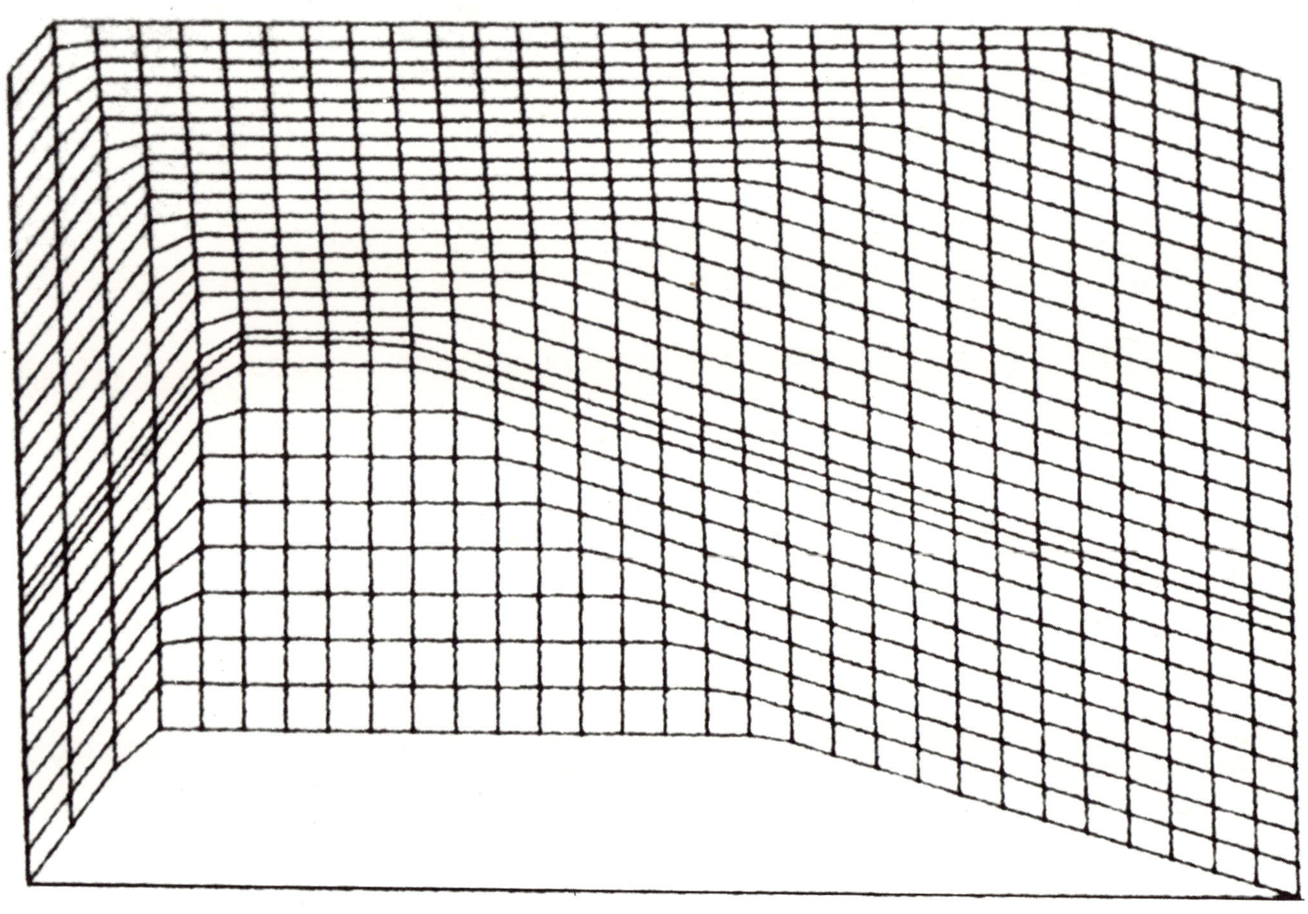

5–40 Summer solstice, 9 AM.

comfortable with the concept, they learned to push at the limits of the envelope and find new form within its discipline. I want you to see for yourself what happened to their designs over a year. It is significant that, from a kind of over-response to the constraint, they learned to break away and finally to do what they wanted to do.

They were helped in their efforts to gain freedom of design choice because in gaining solar access and being assured of that access, a whole range of design tools come into existence that otherwise would not be available to the designer in a normal situation of adjacent buildings. One can site; one can locate; and one can form buildings without being afraid that the energy that one puts into the design of a project is going to be totally negated by the uncertainty of a neighbor.

As the studio work proceeded, important design opportunities began to emerge, with implications for development that may be regulated by future solar zoning. The opportunities included courts and terraces as design elements in response to program requirements for height and daylight. The courts and terraces both tended to concentrate the rhythmic characteristics of the sun's movement, so that one of the most rewarding discoveries of the studio work was that the sun's movement added a time dimension to architecture. The exploration of that dimension then pointed clearly to rhythm as a design strategy.

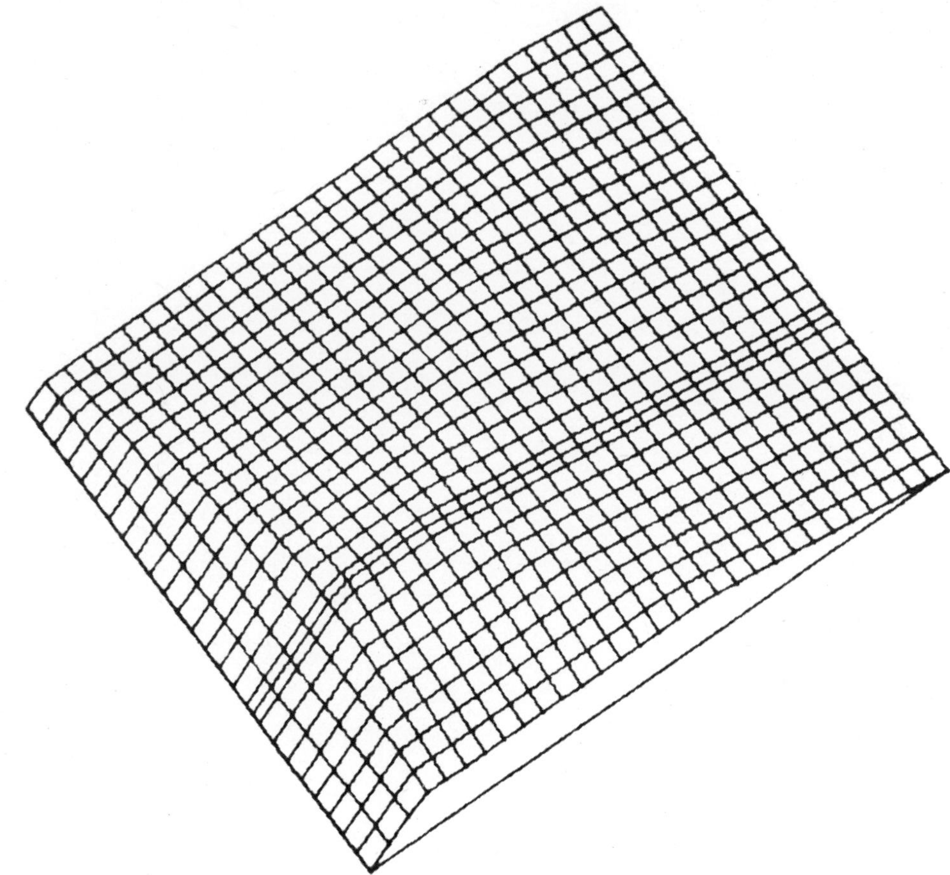

Summer solstice, 11 AM.

5–41

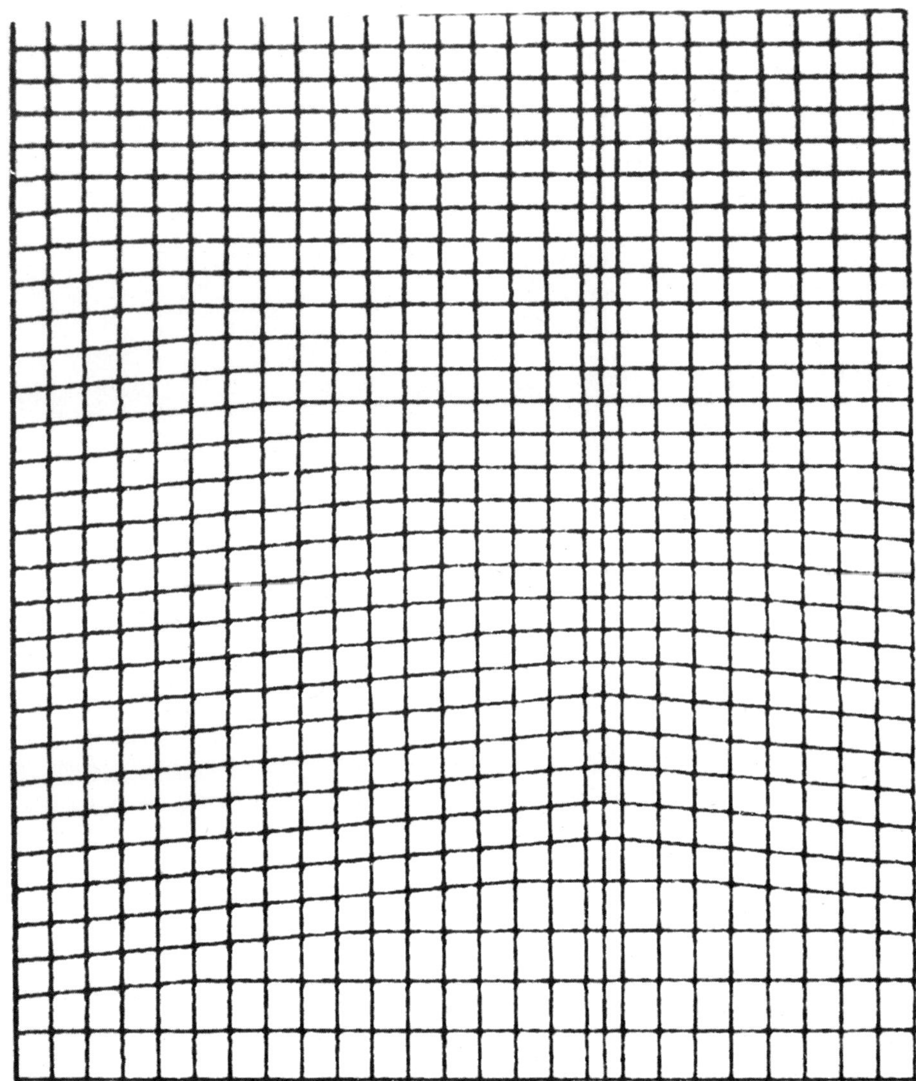

Summer solstice, noon. 5–42

The first project of the fall semester, 1977, was a simple retail store. It was specifically programmed to focus on the solar envelope as an expression of the public's mandate for solar access. The building problem amounted to little more than a shell containing simple retail functions.

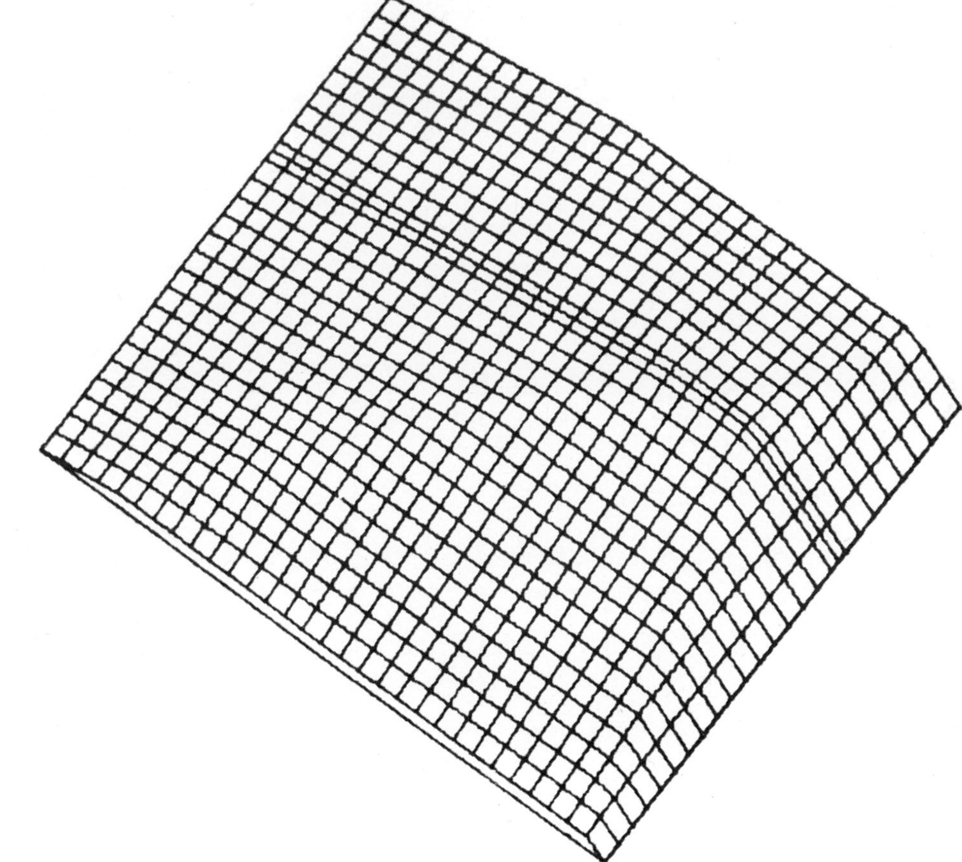

Summer solstice, 1 PM.

5–43

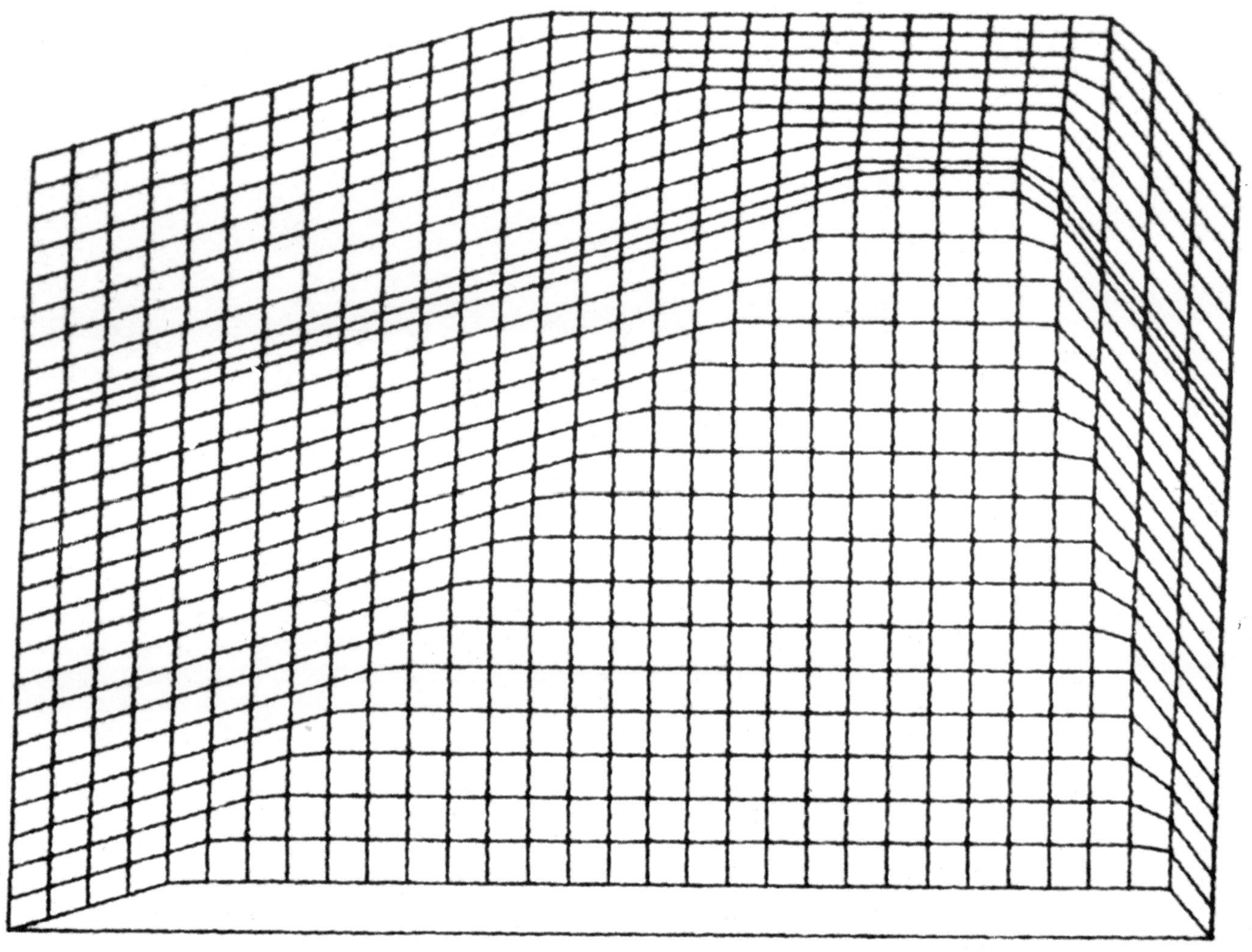

5–44 Summer solstice, 3 PM.

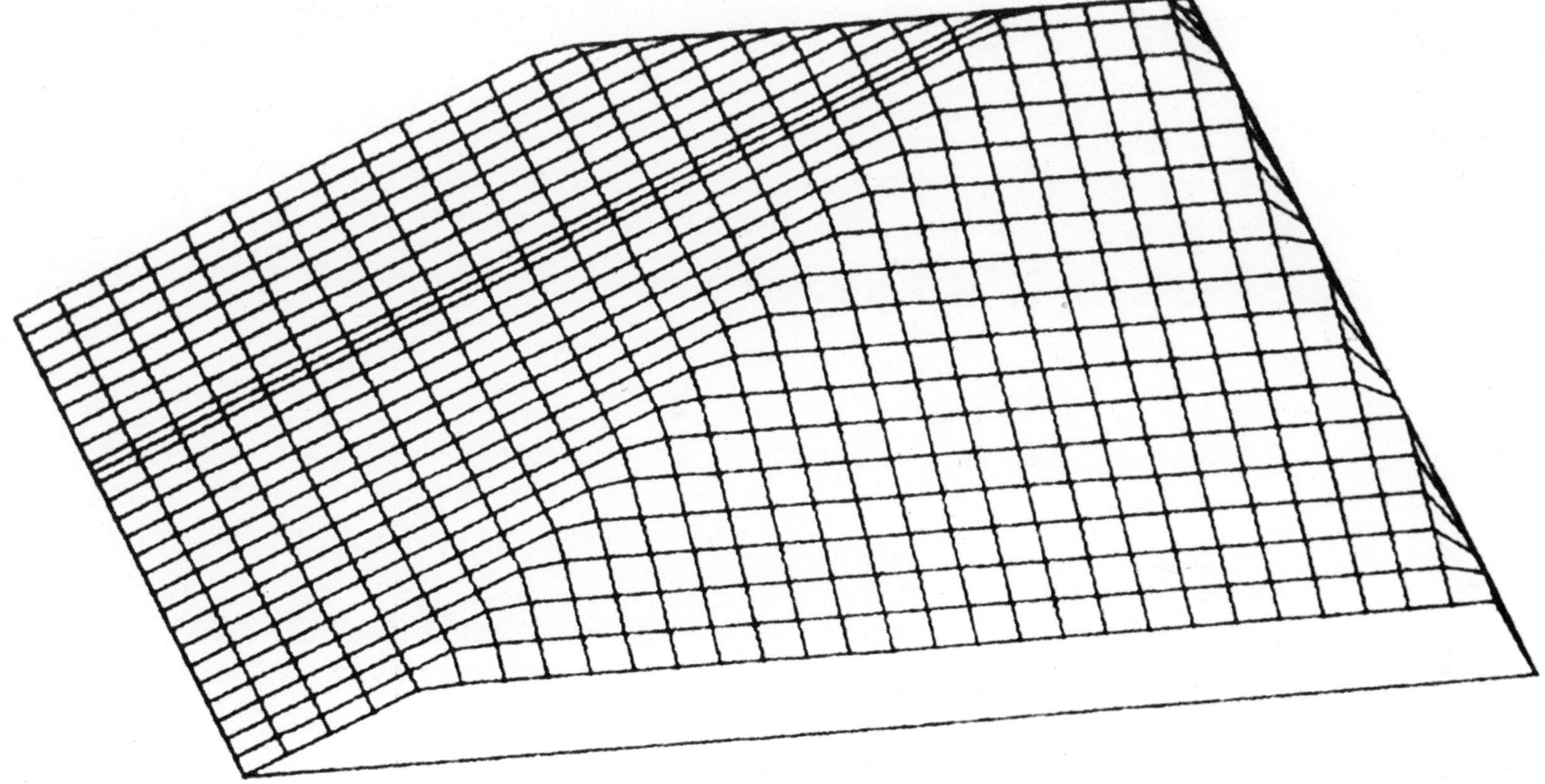

5–45

Summer solstice, 5 PM.

Retail store by Paul Bodine. Viewed from northeast on winter morning. 5–46

The retail store in Figure 5-46 has virtually the same shape as the solar envelope; the building is a literal interpretation of the envelope. North was the only facade in which direct light was allowed into the building. The project shown in Figure 5–47, however, is already starting to depart from the form of the envelope. For example, this designer has pulled the building away from the envelope in order to introduce light into the building in a greater variety of ways. And the store in Figure 5-48 shows even greater freedom. But even this solution is quite constrained by the envelope. It is still a single parameter design.

Retail store by Tom Chessum. Viewed from northwest on winter afternoon. 5–47

Retail store by Armando Caballero. Viewed from northwest on winter afternoon. 5–48

As we went into the semester, the problems became more complex. The second problem was a small office building, which was programmed to focus on a broader design palette than the retail store. Requirements were included for daylight and for more complex uses of location and form as design responses to land use and climate. In Figure 5-49, again there is some concern with filling out the limits of the solar envelope, even some sense of intimidation. But there is also a breaking away in some places. This is the south orientation and the south is beginning to open up to explore the possibilities of using that south sun. In Figure 5-50, we're moving around the corner and seeing the north side of the same building. It is markedly different from the south orientation, which opened

Office by Alan Victor. Viewed from southeast on winter morning.

5-49

up to the sun. Figure 5-51 is a different solution to the same problem, but with a different designer and a different site. Instead of the orientation of the last building, this one faces south on the street so that the front entry, instead of the side entry, is to the street. As you can see in this project, the ridge of the solar envelope runs north-south, yielding broad east and west faces and therefore exposure to daily variations of the sun.

Office by Alan Victor. Viewed from northeast on winter morning. 5-50

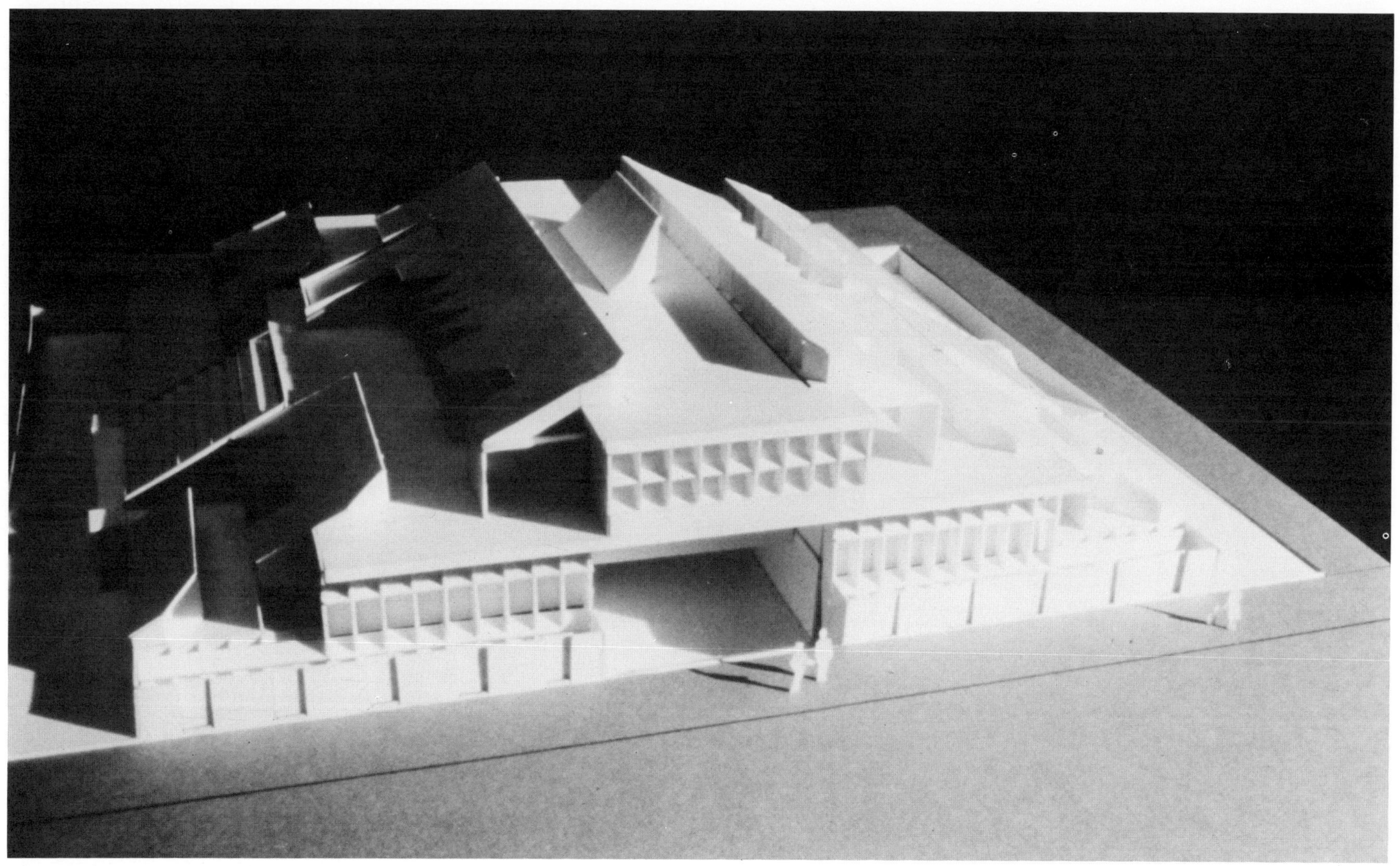

Office by Tom Chessum. Viewed from south on winter morning. 5–51

The problems of allowing light into the building were also becoming more complicated. You just can't close the outside world out of an office building. The solution in Figure 5-52 was to control the light, and here the designer has begun to use the space under the envelope to control light and to enrich the space, rather than simply being concerned with literally filling the space.

The third studio problem was not only a small office building, but a government center containing a number of functions, including a library and a small auditorium. The solutions became even more differentiated. Buildings were juxtaposed, oriented and related to the surroundings to introduce useful light and heat into courtyards and interiors. Their shapes and structures were adapted for the same reasons. A significant diversity of built forms emerged and buildings clearly varied from one orientation to another.

In all of this work, we were concerned with what could have been built under present zoning in Los Angeles. Our aim was to keep increasing density, pushing at the limits of the solar envelope to build a strong development case for solar zoning. All of the normal problems of a building program had to be resolved; we wanted good buildings by all conventional architectural criteria. But we tried to solve these problems within the context of solar zoning.

Office by Tom Chessum. Viewed from southwest on summer morning. 5–52

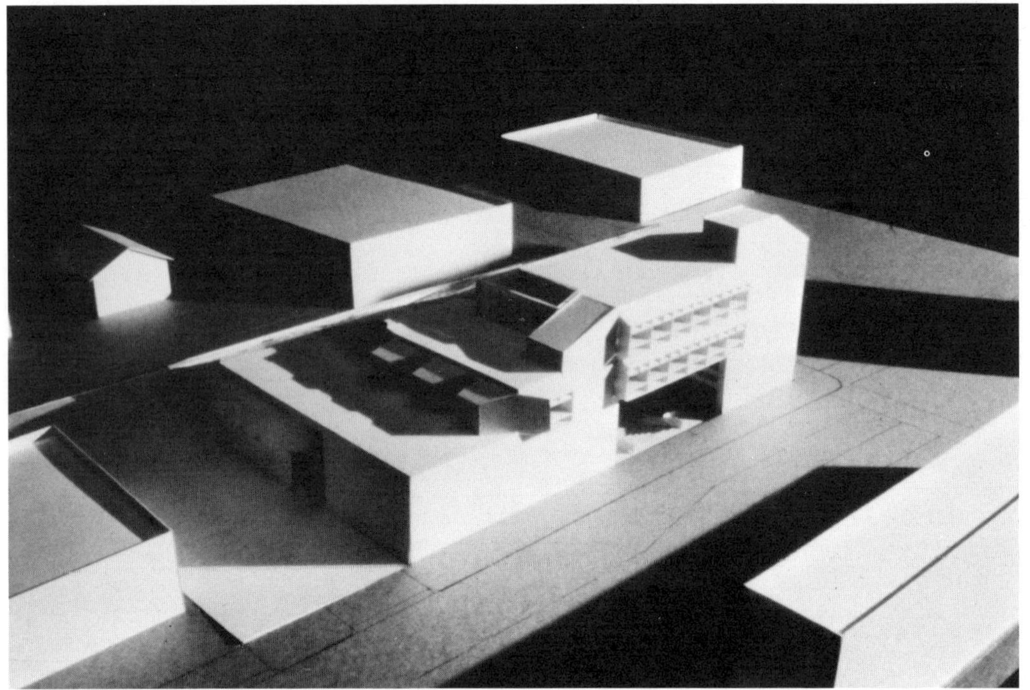

Government center by Nora Sarkissian. Viewed from southwest on winter morning. 5–53

The government center project shown in Figure 5–53 had a corner site, rather than a site in the middle of a block. The solar envelope can therefore extend across the street on the south and east. This not only enriches the form possibilities within the envelope, but it increases the developable volume. This is the south orientation and the entry to the building. Here one can begin to see differentiation between modes of sun control. One side is different from another side, and there are interesting transitions around corners. The diversity and variety suggest designer choice.

In Figure 5–54, we are looking at the same project from the west, and the long slope of the north facade becomes evident. In this case, the library is located under the north slope and light is allowed inside. But this building, with all of its diversity, does not cast shadows on its neighbors during critical time periods. Sometimes its neighbors cast shadows on it, but the designer was responsible for adapting to that circumstance. Buildings tend to last a long time, so it is clearly a responsibility of a designer to adapt to existing conditions.

Our next design problem was housing, which was programmed to focus on the human dimensions of environmental quality. Its smaller scale seems to expose more issues of design quality, and the students felt even more free to break away from the envelope's domination. Perhaps some of this freedom was derived from a greater familiarity, and therefore comfort, with the building type, or perhaps the scale is closer to human proportions. In any case, the correlation of building form with natural variation resulted in greater architectural diversity and choice.

Government center by Nora Sarkissian. Viewed from west at equinox noon. 5-54

In Figure 5–55, the solar envelope is for a 100-foot frontage lot running long in the north–south direction. The faces of the envelope are therefore oriented east and west and subject to a daily rhythm of solar impact. In addition, local zoning allows a 6-foot fence around a residential property for privacy. This effectively lifts the envelope, and expands the amount of developable volume within it. Given these constraints, the designer has chosen to exercise the freedom to locate as a mode of adaptation and to choose a building orientation. Beyond that, this project exhibits form differentiation responsive to orientation. The building closes down on the north and opens up on the south.

In Figure 5–56, we can begin to sense a polarity in the design. We can see that this design is highly differentiated on the north and south in response to seasonal variations. The building is less sensitive to daily variations. This reflects a concentration on seasonal rhythms in the early studio work. In Figure 5–57, however, you can see the north slope and its tendency to close, letting light inside in sometimes interesting ways—bouncing the light in, for example, but in general opening up on the south and closing down on the north. This, incidentally, is the model seen on a summer day, and the sun does not penetrate the building facades.

In winter the low winter sun penetrates the building deeply. The designer responds to these seasonal pulses by differentiating the form and function of the building.

Housing by Rich Knarr. Viewed from northwest on winter morning.

5–55

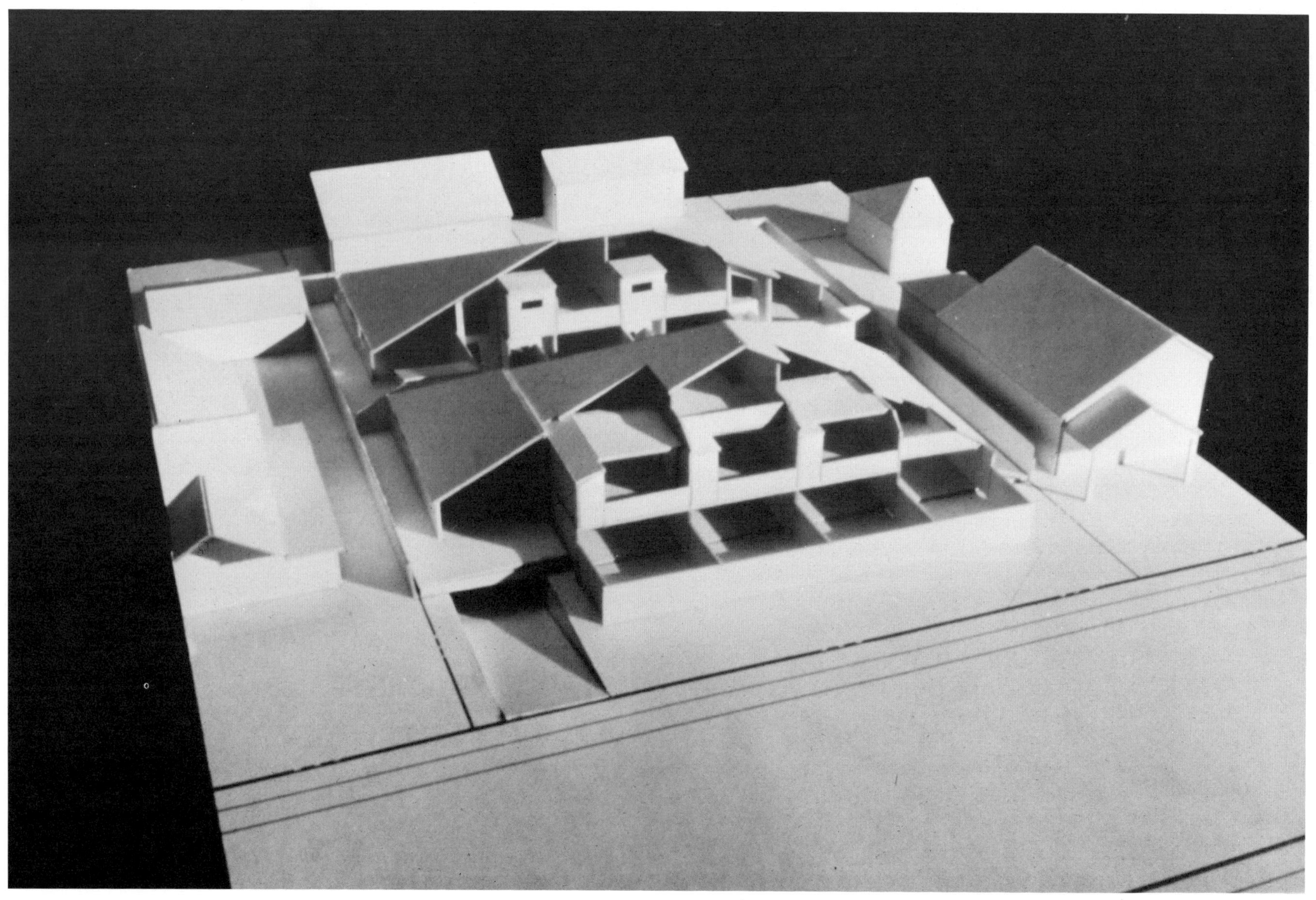

Housing by Rich Knarr. Viewed from southwest on winter morning.

5-56

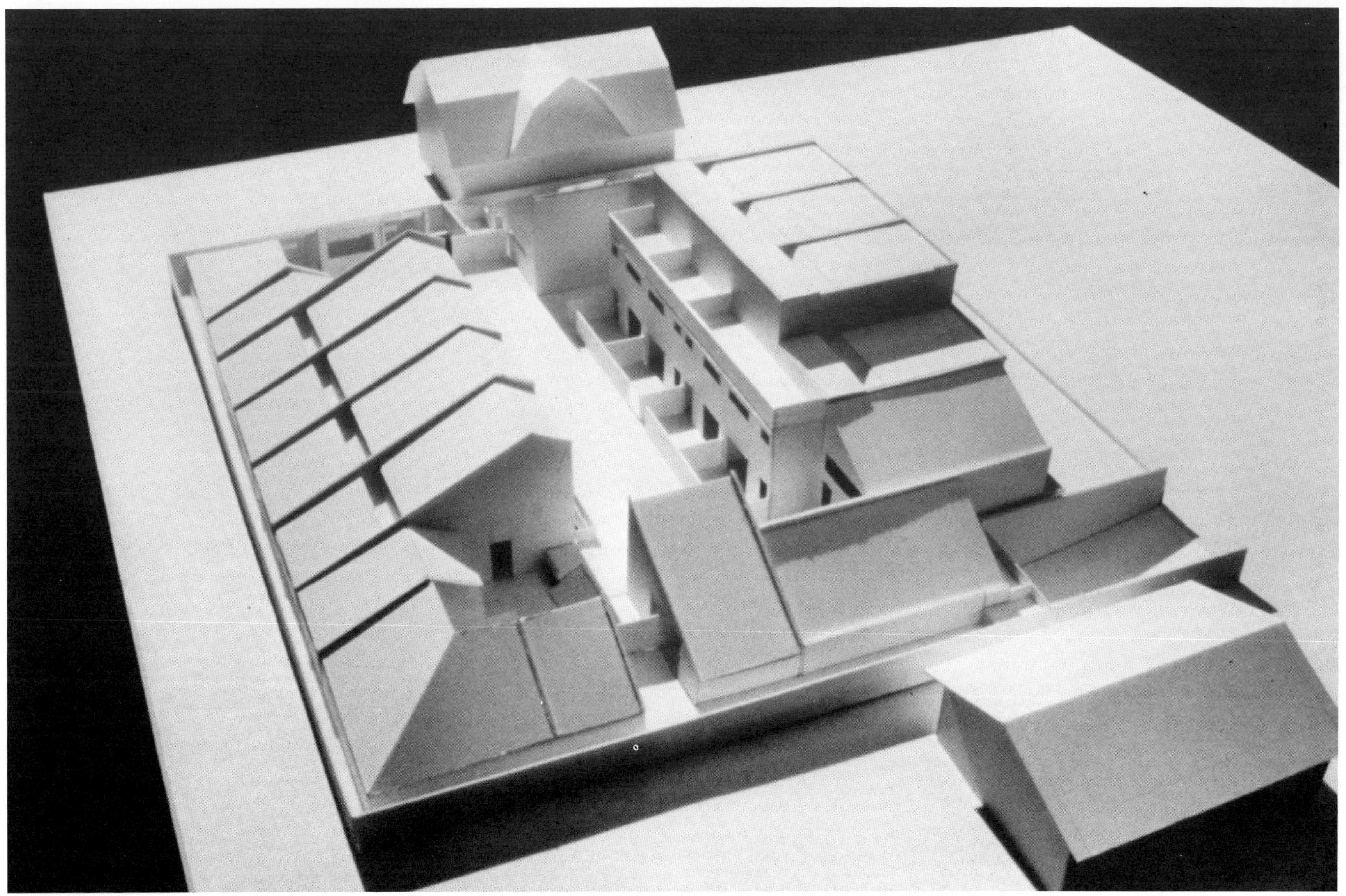

Housing by Sarah Dennison. Viewed from northwest on summer morning. 5–57

Housing by Paula Waller. Viewed from south on winter morning. 5-58

The designer of the project shown in Figure 5-58 takes a different approach. Her solar envelope was long in the north-south direction, with its faces to the east and west and therefore subject to daily variations. She has a high ridge and a long ridge to deal with. She was presented with the choice of greater density, for which she could fill the envelope and deal with the east-west orientation by design. Or, she could choose greater south exposure and change the building's polarity from that of the envelope. This, in effect, substitutes two smaller design envelopes for the basic site envelope and gains south exposure at the cost of density.

The idea of reorienting development to obtain a double tier of housing, both facing south, did not seem economical to the designer of the housing in Figure 5-59. Instead, she chose to fill the envelope to the ridge, and to work within the east and west exposures, differentiating them for sun control. Where she had south exposure, she used it. The units that face south are designed to capture the winter sun and exclude the summer sun. The ridge units are separated by party walls, on which light from skylights can play, allowing the capture of both heat and light.

She dealt with the east and west faces by not treating them the same way (Figure 5-59). She brought the glass out farther on the east, producing a completely different quality of light and heat. In this illustration, you are moving from the east, around to the north, which is quite closed down and also happens to face the street. It provides a kind of privacy. The glass on the west facade was set deeper than that on the east, so that there was a spatial correlation with the morning and afternoon solar dynamic. One can still see the overall imprint of the envelope, but within it there is a lot of departure.

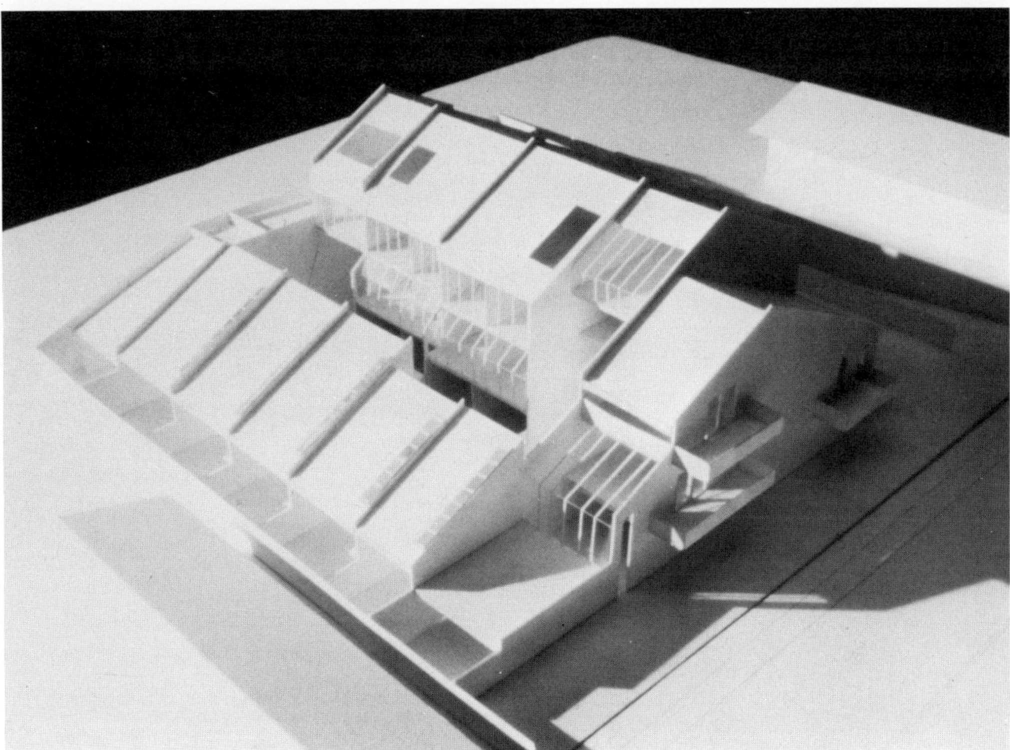

Housing by Paula Waller. Viewed from northeast on winter morning. 5-59

Figure 5-60 is another example of housing design that clearly distinguishes between north and south, so that the seasonal migrations of the sun tend to intensify the differences and to intensify the form itself. The distinctions, diversification and differential between one orientation and another become much more intense. And this project adds a new element, not just that of sculpting form by light, but that of time, and not just a change in time, but a rhythmic change that correlates with natural variation.

Housing by Inez Gomez. Viewed from east on winter morning. 5-60

Clear, contrasting qualities appeared on the north and south facades of these designs (Figure 5-61). The south opened up to the sun; spaces were larger, higher and deeper; and the building's structure was more complex. The southern row of housing was generally much higher than the northern row, with north-south development quite asymmetrical.

Housing by Inez Gomez. Viewed from southwest on winter morning. 5-61

In Figure 5–62, it can be seen that the north side is simpler and more enclosed. Spaces are smaller, lower and more shallow to cut heat losses and make use of whatever light penetrated from that side.

A final point about this project can be summarized in a phrase. Location determines rhythm; rhythm determines form. This implied sequence in the phrase was employed as a strategy by the student designers to achieve comfort and choice in

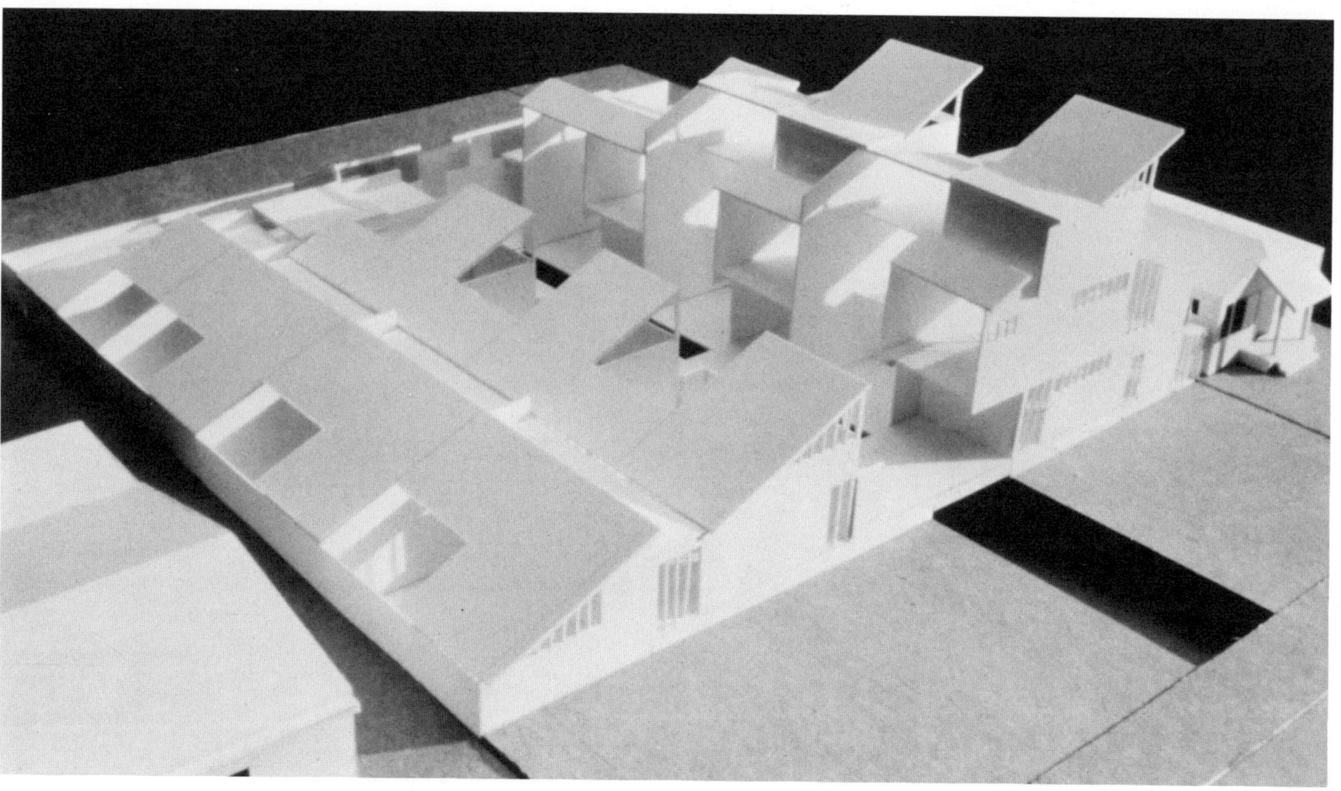

Housing by Inez Gomez. Viewed from northwest on winter afternoon.

5–62

their housing designs. This strategy could be clearly seen at work on an east-facing site that employed the orientation and juxtaposition of two rows of houses to gain southern exposure and to accentuate the seasonal advantage of low winter and high summer sun (Figure 5–63). The location of the buildings determined their basic rhythm. The seasonal rhythm, in turn, required a form adaptation to take advantage of the southern exposure.

Housing by Tom Chessum. Viewed from northeast on winter morning. 5–63

The relative size, shape and structure of the buildings in Figure 5–64 was chosen by the designer to make the best use of the solar dynamic to mitigate seasonal variation. The highest portions of the buildings were specially treated to guarantee solar access on the north. Sometimes this meant diagonally cutting shapes in two directions to avoid shadowing within the development.

Functional differentiation within the units also occurred. When the rows were close together, the upper units naturally received greater access to the sun. Major living spaces were located above minor ones; living and bedroom areas were sometimes on top, with kitchens and dining rooms below. Thus, a kind of vertical migration evolved within the building, based on solar access.

Housing by Tom Chessum. Viewed from southeast on winter morning.

5–64

Figure 5-65 is a similar view of the same building, but taken on a winter afternoon instead of in the morning. The east elevation is dark in the afternoon, but receives the full impact of the morning sun, when ambient temperatures are just starting to rise. The quality of light in the afternoon is completely different from the quality of light in the morning.

Housing by Tom Chessum. Viewed from southeast on winter afternoon. 5-65

There is also a kind of scaling of the impact of the new buildings on the existing bungalow adjacent to them. The solar envelope starts to develop a sense of scale for development that is proportional to the volume of development and to the street.

In summary, building within this kind of constraint emphasizes the sense of natural rhythm, giving the possibility not only of rhythm as a design strategy, but of rhythm as a basis for orientation. By orientation, I am not only referring to orientation in space, but in real time.

The urban design implications of orientation are significant. One block of a city may not develop the same grain as another block of opposite orientation, especially as cities rebuild, as many are doing today. We may choose to leave some orientations of the grid alone, and only to modify others for purposes of energy conversion and life quality, establishing a correlation between grain size and orientation of city blocks. Thus, in Lynch's terms, we would have orientation in relation to the city as a whole, based on (as Parr says) our perceptions and based on (as Fitch would say) a concern for energy.

But for all of us, in whatever terms, there is a concern for life quality as a function of a clear basis in space and time for choice, and for choice as a measure of life quality. It was Mumford who said, in a marvelous little book called *Art and Technics,* that "choosing is a creative act."

To choose is to create and, to the extent that making an artificial system in balanced energy response to the main recurring forces of nature produces a diversification of that system, solar design provides choice. That is the extent to which balanced energy approaches enhance life quality. If choosing is a creative act, then what we are all about in taking an energy-conscious view of buildings is providing an environment in which people can act more creatively. I would submit that this in itself is a noble concern for any designer, whether energy concerns the designer or not. The issue of the quality of the environment is—has to be—of concern. To the extent that this kind of approach enriches and diversifies the environment, it enhances choice and suggests that energy and life quality are not incompatible objectives. They may, in fact, be seen as intertwined and supportive. To do one is, in fact, to do the other.

6. Environmental Performance

Cesar Pelli

I am not particularly knowledgeable about energy technologies, although I am interested in them. One reason I decided to contribute to this symposium is that I could learn much from the other contributors. But another reason was that my own concerns may be useful to you, as they are views from a different perspective. I am primarily an architect and a designer. Eventually buildings get built and they have to be built with some idea, not only about energy, but about a number of other issues, some of which interest me very much. I have always been concerned with the way in which buildings consume resources. By that I mean all resources—aluminum, brick, etc., as well as energy or time. I have a predilection for achieving architectural design ends with the greatest economy of means.

But one also needs attitudes about aesthetics; this may, in fact, be the key issue, because we build what we *like* more often than we build what we *should.* This is a very good moment to be discussing issues of design as it relates to energy, because a range of new imperatives are emerging. We need aesthetical theories responsive to our social needs.

It has been said that the modern movement is dead, and has been replaced by new movements with a variety of fancy names. I don't believe it. It is not a matter of movements, of changes in fashion, but something deeper. Because this is a moment of fundamental change in our preconceptions about design, we need good models to help us deal with the situation. Architects cannot make decisions about architectural form without some concept in mind prior to those decisions; and while design concepts are partly a matter of individual imagination and personal preference, they are also a product of a collective consciousness. Theories give explicit expression to that consciousness. The present lack of theory—"anything goes"—is both stimulating at one level and seriously inadequate at another. We need, collectively, to once again take positions on what is "good" and what is "bad."

I would like to talk about all this in relation to my own practice. Certain concepts come to your mind which you realize you have been adhering to for years without knowing it. One of these is the concept of performance in architecture. The word "performance" is a very good word; it hasn't been used much in discussions about architecture and therefore is still fresh and capable of carrying real meaning, unlike words such as "modern" and "functionalism," which are old, tired words no longer able to make us think. The word "performance" means many things that are useful to keep in mind while looking at my work. First of all, performance refers to the completed action—that is, the completed object, something that is finished. It does not refer only to process, but to process *and* the finished project. It also has another meaning, which is the concept of theatrical performance, or drama. This is a powerful analogy for architecture: buildings are the setting within which the drama of life is played out. A third meaning comes from engineering. It measures the efficiency of any mechanism or construction and therefore implies that there is a way of testing architectural end products. I believe that architecture should be subjected to tests because I believe that architects are responsible not only to themselves but to society. And indeed, society itself believes that very much; this accounts for the proliferation of codes, regulations and lawsuits. Society makes the architect not only responsible, but *accountable.* Therefore, it is time to start measuring our buildings by their performance.

The concept of performance is very different from that of functionalism. Functionalism is only one aspect of performance: the compliance with a pre-existing program. It implies a closed system. When one says, "form follows function," it is implied that the function pre-exists, for the form to follow afterwards. In reality, during the process of their planning and erection, buildings to a large extent create their own functions, resulting in much more complex systems than any program can dictate. On the other hand, performance relates not only to compliance with a program, but also, and more important, to the way that a building actually behaves, to the way it is perceived by the users, to the way it responds to surrounding buildings, to how it relates to climate and to the way people actually feel and experience its spaces; in short, to all of the aspects of what the building is.

We are still confronted with many of the same questions that were raised by the polemicists of the early modern movement. They made the observation that a great change had occurred in what buildings were, and they thought that this change was definitive and revolutionary in character. Some of these observations were inaccurate and many of the conclusions and remedies were exaggerated or off the mark. Nonetheless, there have been fundamental changes that are, for the foreseeable future, irreversible. Consider changes, for example, in technology—and I don't mean just the technology of construction, which is secondary, but basic technology like the internal combustion engine or indoor plumbing. Just think how differently a building is designed with or without plumbing; or electricity; or elevators.

There are also changes, of course, in the basic technology of building. The most fundamental one occurred in response to social rather than material forces. Masonry buildings derive their formal qualities from methods of stone construction. They depend on one essential thing for their existence: the availability in great abundance of cheap, highly skilled labor. I read recently some letters of H. H. Richardson on the construction of Trinity Church, and the description of the stone quarrying and stone shaping reminded me of the description of the building of some of the early gothic cathedrals, which was probably not too different from the way Karnak was built. This is archaeological speculation, but the techniques of stone construction remained basically unchanged for over three or four thousand years.

Figure 6–1 is from the recent Museum of Modern

Art exhibition of drawings from L'Ecole des Beaux Arts. It indicates a preoccupation with, or a desire to return to, forms and attitudes about architecture which had been considered abandoned. The architectural elements in the drawing are related to one particular quality of older forms, which, because of the universality of stone construction, reflect the need to distribute a heavy weight downwards. The form of the Ionic column is, among other things, an expression of weight transfer (Figure 6-2). The shape of the capital tells us how heavy the weight is and how it is distributed from the stone lintel to the vertical shaft. The base mediates and transfers the weight from the vertical shaft of the column to the horizontal ground. Even when we talk of covering space, the forms which had meaning in the western tradition depended primarily on the characteristics of stone construction.

6-1. Drawing from L'Ecole des Beaux Arts

6–2

Because stone construction has lost its central position in western architecture, the system of forms derived from it cannot have real vitality and it will not help us solve the new and critical problems of our times. It is essential that we develop an aesthetic system based on what we have and what we need to accomplish. These concerns are always present in my design, and my thoughts and insights may be useful to you.

My interest in pre-masonry buildings is recent, so that I have not a theory, but only some impressions: they have a quality of forms that is quite different from the quality of forms in buildings derived from a tradition of stone construction.

Figure 6–3 is interesting because it is of a very early, and truly American building. It is minimal. The frame is independent of the skin, while the parts are made elsewhere before they are assembled. There is a fire inside, indicating a close relationship between the shelter and the energy source. That goes back to the beginning of architecture. Buildings like this, are found all over the world. They are very beautiful. They are primarily silhouettes; aesthetically, they depend on the expression of the total form, not of the parts.

Huts and tents are probably the most democratic buildings ever built, in the sense that the people who commissioned them were the same people that built them. I've always thought that the best way to measure the degree of power-sharing in a society is to look at the number of people able to commission a building. The more people that can commission a building, the more democratic or the greater the economic participation in the system. It doesn't matter if it is through a board or a committee; and if you can commission your own house, you are achieving a high degree of participation in the collective power of that society.

6–3

6–4

Figure 6–4 shows a single-family dwelling. The family itself builds it, so the correspondence is almost one to one. Look at the absence of modulations in the form. The form is beautifully total. This quality may be found in indigenous architecture in many parts of the world, from Brasil to Mongolia. Surface decoration has a two-dimensional, graphic quality.

Black tents (Figure 6–5) are among the most sophisticated types of shelter ever developed. It is a familiar form of habitation all through North Africa, the Mid-East, Afghanistan and northern India. Again, the form is total. The qualities of form in this type of building are very different from that of masonry construction, which only started, at the earliest, about 10,000 years ago, and probably became the most common type of shelter only about 400 to 500 years ago. Until then, most people in the world lived in tents or huts. So the tradition of masonry construction is really very short. We have been living in variations of this type of lightweight shelter for hundreds of thousands of years, and they are deeply rooted in our culture.

One of the best known icons of the modern movement is Mies Van der Rohe's design for an office building in Friedrichstrasse, Berlin 1919 (Figure 6–6). Its form is direct and total as in a tent or a hut. In this early design, Mies displayed a very good understanding of the expressive possibilities of a glass envelope architecture.

6–5

6–6

Figure 6–7, on the other hand, is an example of weight-bearing construction: Mendelsohn's Einstein Tower in Potsdam. Although it, too, is an important modern movement icon, its formal qualities are quite different from those of a skin enclosure architecture. There is light and shadow, there is modeling, there is a rich particularization of the elements of the form. In this case, the achievement of these particular formal ends was of paramount importance to Mendelsohn; the actual means employed were secondary. It is brick masonry, rendered over and painted.

6–7

6–8

This brings me to one of my buildings, the Pacific Design Center in Los Angeles (Figure 6–8). It has a very thin glass skin. Only 6 percent of the envelope consists of transparent glass providing view out; the rest of the glass wall is opaque and contains as much insulation as we could convince the owner to pay for. Transparent glass areas are primarily concentrated on the escalator concourse. This is built as a half cylinder, yet because of the reflection— particularly at night—we see it as a complete cylinder, recreating on the exterior the reality of the interior space.

The expression of a skin enclosure often depends on a small detail to separate it from the fundamentally different expression of a frame. For instance, consider the relationship between a mullion and the glass that surrounds it (Figure 6–9). I have used primarily two types of mullions in my buildings; they are quite narrow and project only a little from the glass. The depth of the projection is doubled by reflection. Mies Van der Rohe's detail for the Toronto Dominion Center involves a projection of about 10 inches; when reflected, it is 20 inches. Because of this apparent depth in the facade, the glass envelope disappears and the frame dominates. The aesthetic intention in Mies' design has shifted completely from the concept of enclosing space to the concept of expressing a structural truth. It is important to be aware that Mies knew exactly what he was doing. From early interests in perception, his concerns shifted to a primary interest in intellectual order.

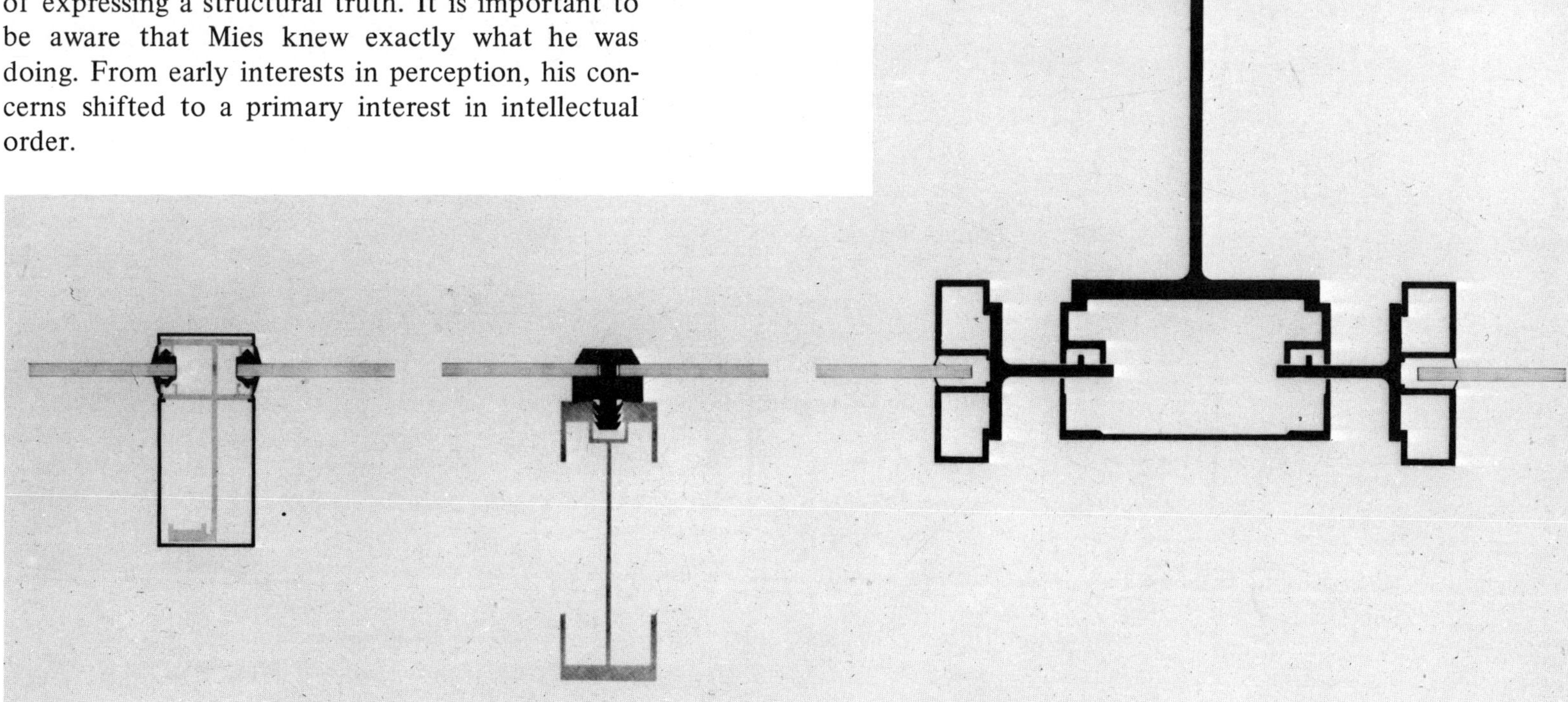

6–9

6–10

The curtain wall used on the Pacific Design Center is probably the type of enclosure for large buildings that has the highest efficiency with the least amount of material, for the least cost within present available technologies (Figures 6–10 and 6–11). It is important to make the best out of what is available. As a design attitude, I much prefer to make the best of available means rather than to distort those means to conform to some preconceived formal end. I think the imperatives for skin-wrapped frame buildings will persist for many years to come. However, this is only true of buildings above a certain height and volume. For buildings of one, two and three stories and restricted volumes, the wood frame will probably remain as the best available technology.

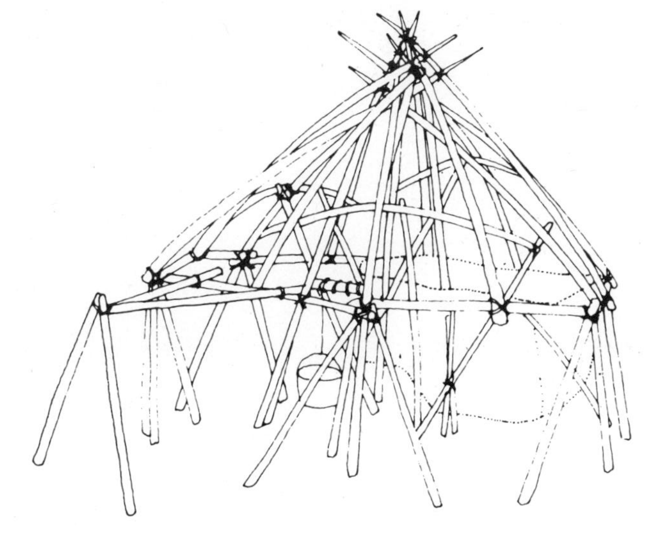

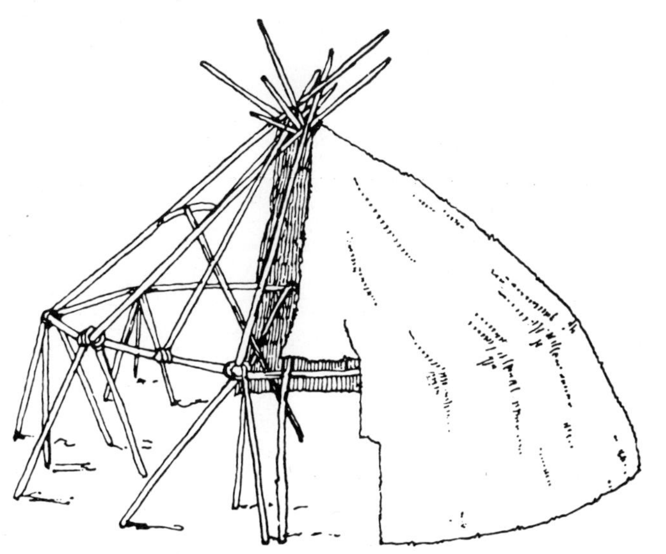

6–11

6–12

Here I would like to say something about scale. It is often assumed that scaleless buildings, such as Pacific Design Center, dwarf people, making them seem insignificant. The truth is that against such a neutral background, the first thing you see is the individual (Figure 6–12). The surface of the building is so simple that any change of patterns, as that of people walking by, dominates. The whole surface feels very light, and people do not feel dwarfed. I sometimes show slides of people walking among columns of neo-classical buildings in Washington, D.C., with tall, five-story high columns, and in these the human figure is totally dwarfed by the mass and weight of the buildings. Here, instead, the individual remains the center of the action.

San Bernardino City Hall is another building we designed as a space enclosure. It is a design approach that, because of the use of this particular skin, can lead to a variety of forms, and respond to many conditions.

San Bernardino, where the building in Figures 6–13 and 6–14 is located, has a very high radiant temperature this was all before the urgent need for energy conservation—and our actions were in response to a situation that we wanted to make more humane, more economical, with a better building, and not because we thought that saving energy was particularly important. The southface almost has no windows and contains all of the services—stairways, elevators, toilets, mechanical shafts, etc. We even have two rows of very deep balconies to bring in natural light to the central corridors. The windows are on the north, with full glass and with little vision glass on the east and west sides. We used insulated spandrel glass in conjunction with transparent glass to achieve a visually uniform surface within which we were free to handle enclosures in different ways, as requirements dictated. The aesthetical intentions were made to coincide with our social and pragmatic objectives.

6–13

CITY ATOR
FINANCE
ADMIN
ACCOUNTING
CITY
ADMINISTRATION
OFFICE
OF THE MAYOR

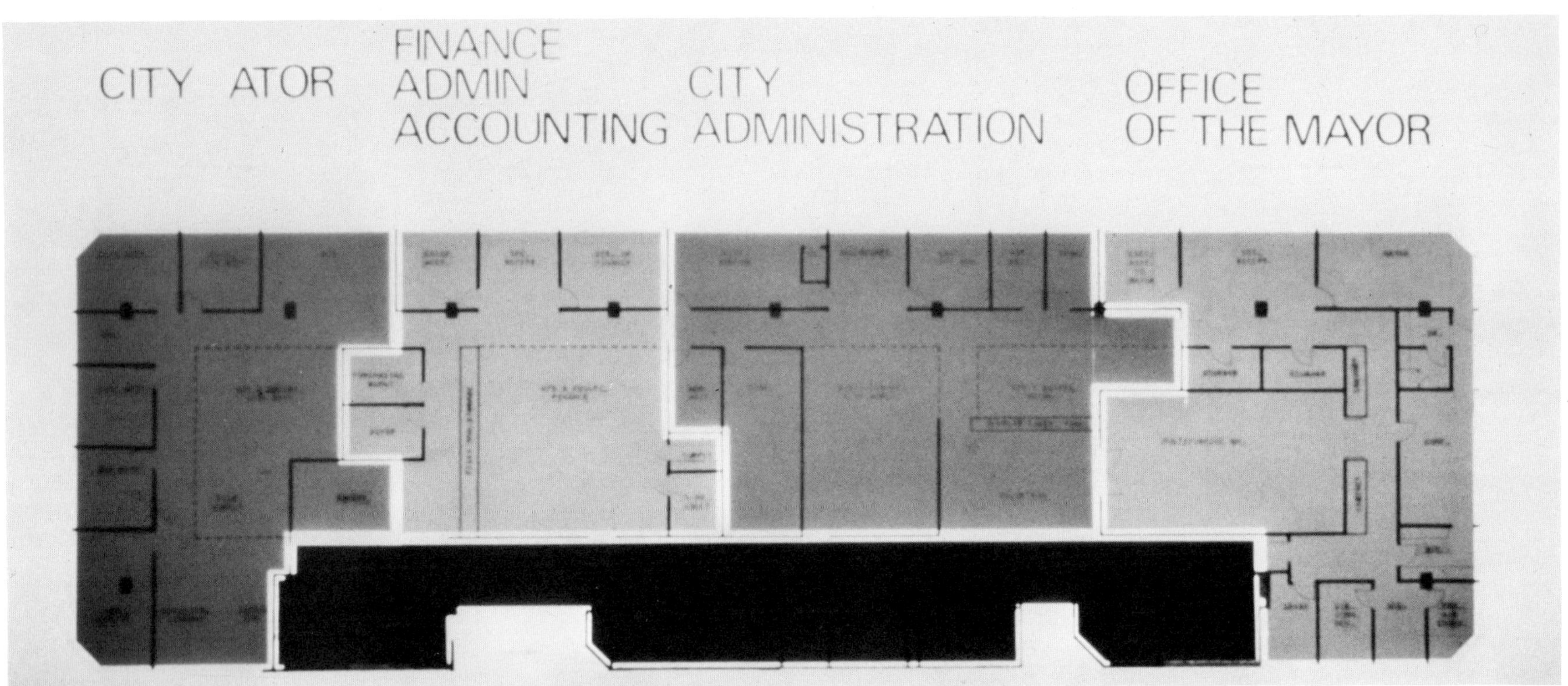

6–14

Figure 6–15 illustrates another building that deals with similar problems—problems that are, like all architectural problems, a conflicting combination of practical, aesthetic and theoretical issues that one tries to resolve in one single solution. This is an office building in Oakland, California, made of metal and glass. By using stamped sheet metal for the spandrels, we discovered that we could get an extraordinarily cheap wall at that time. We knew of a firm that had bought an old metal stamping press from an automobile company and they were doing imitation precast and stone facades, so we used it to make the wall look like stamped metal. The windows on the south facade are very narrow. They go from sill to door height. They are larger on the east and west, and much larger on the north. When we first proposed this to our client, they were quite worried by an office building with equal rental values on all sides that would look different on each of these sides. But afterwards they liked it very much, and now, in their brochures, they describe the building as a tree trunk covered with moss on the north side.

I have been very interested in different perceptions of architecture and architectural form. Le Corbusier defined architecture as "the magnificent play of light and shadow on a volume." He was, of course, referring to the western masonry tradition. I could see that glass buildings do not do that at all; they go very dead in sunlight, and are actually very beautiful when in the shade, where the reflections are stronger. As seen in Figure 6–16, I tried to reconcile both expressive modes. Normally, you cannot have both at the same time, because if you make strong modeled forms, and then fill in

6–15

6-16

the spaces in between with glass, the projections of the forms will completely destroy any reflections. But by making the forms curve inside the plane of the glass, they do not obstruct the reflections. This means that the rules are reversed. The ground becomes figure and the figure becomes ground, as you go from light to shadow. When there are no reflections to give interest, you still have the shiny glass reflecting the sky, which is never uniform.

The skin wall to which I have been referring is made up of sheet metal, pressed and sprayed with fluorocarbon paint, with as much insulation inside as is desirable. Combined with the glass windows, it has great visual richness, if you take into account and plan for not only the permanent elements, but also for the changing elements, of the building, the elements that are not controllable by the architect, and that requires, again, a new way of understanding buildings. It means that the most important aspects of your buildings may be outside of your control. I find that quite extraordinary and beautiful, although some architects get very nervous with this idea.

This building for COMSAT in Maryland (Figures 6-17 & 6-18) has a simple organizational system. In this case, the concept of performance took on a special meaning because of the particular character of the building type. Complexes have become very prevalent in the industrialized world. They are collections of large and discrete functions. When integrated within a single formal entity, buildings result in a magnitude quite unknown in the past. My approach to a resolution of this design problem was to use a clear and generous circulation concourse to link the different entities. We thought that this was the most important element in the building, because it was the only element that was common to everybody, from the president of the company to the office boys. We considered that it should be like an enclosed main street, with the best views and two-story ceilings. To our surprise, it was accepted by the company. When we designed this project, in 1968, we thought that it would be appropriate for a laboratory building involved in high technology to have a technological expression: curved aluminum panels with inserted mirror glass windows. Very narrow external corridors provide views out, and, when looking at the building, we see not only the exterior, but also through to the internal areas of the building and beyond into the space of the courtyard. If this had been a masonry building, all that we would be able to see would be the exterior form. The building is a man-made object sitting in nature, in mutual respect. The metal curves in all directions to imply and strengthen the idea that this is a skin independent from a structure. The metal provided some delightful surprises; for example, the way it picks up the color of the sky.

6–17

6–18

A project that is particularly dear to me is shown in Figure 6–19: 830 units of public housing in Hawaii, Kukui Gardens. We had to re-examine many of our values and predilections. Response to climate proved to be a primary necessity. The units have no mechanical systems whatsoever—no heating, no cooling. In Hawaii, this is possible, if one opens up the units to the breezes and all the units were designed to have cross-ventilation. Another important issue in this project was cost. The project was financed by a government subsidy in the form of a low-interest loan. The cost of the building reflects directly in the rent paid by the inhabitants. As designer, I was in a position to save money for those who would be paying rent. Saving $10 in rent for someone who was making $6000 or $7000 a year was, for me, much more important than any aesthetic refinement.

Yet another important thing we could provide was a quality of decency. By decent I mean something the middle class would enjoy living in. We tried to achieve this in a number of ways. We made the units easily accessible. There is a one-story unit that is entered directly at ground level, and then on top of that is a two-story unit, so there is only one floor up to reach the higher units. No one has to climb three flights of stairs with heavy groceries. We developed a very simple structural system consisting of concrete block bearing walls and precast planks. Each unit has much of the character of a house. Each has one private exterior space, at least the size of a room; a backyard for the lower units; and a lanai in the upper stories. In the Hawaiian climate, such an exterior space becomes an extra room for very little money. The units were organized around a park with perimeter parking and pedestrian movement toward the center. We very much wanted to allow delivery vehicles into the side streets, but concepts of separation of pedestrians and vehicular traffic on the part of the planning authorities unfortunately prevented this. I believe they would have enriched the life of those little streets. The units are modulated so as to provide not only streets which are places of gathering, but also public places for each neighborhood. Each one has a special function and has been designed differently, but all are places for groups to gather in order to foster a sense of community. The best thing I heard about these units was from a taxi driver who started arguing against the government, saying that it was a terrible government that only protects the very rich and the very poor, and does nothing for the people who work. He said that these housing units for the poor allowed them to live better than he could. Here was someone in the middle class finding this a desirable place to live.

6–19

6-20

Figure 6–20 shows another housing project, in the Santa Monica hills, representing a different scale of values. Our main concern was to preserve the profile of the hills. If it was necessary to build that number of units, what was the best way to place them? Our answer was to make them as much a part of the hill as possible, and not build apartment towers as had been suggested.

In the next two projects (Figures 6–21 and 6–22), the design solutions arose directly from accepting the very tight requirements demanded by our clients. The first is a post office in Los Angeles (Figure 6–21), where we accepted the bell column and flat slab system employed by the U. S. Post Office system in their projects. All we did was to design the exterior wall. We cut the conical capitals two inches off the face of the column (this gave us a hyperbola) and we infilled the space between columns with brick, very tightly. With these simple decisions, we achieved a high degree of dignity in a very economical building.

In the second project, the American Embassy in Tokyo (Figure 6–22) we had to work within stringent State Department requirements. The building is simple and direct, both in its structural system and in its architectural expression, which means it was inexpensive to construct. The structure is totally uninterrupted throughout. It was required that the exterior wall should be of concrete; that there would be a maximum glazed area of 25 percent; and that all windows should be openable. The organization is extremely simple, and divides the plan into two very thin layers. Each layer is one office deep; so if you have an office with a secretary's office next to yours,

6–21

6–22

he/she can also have windows. The courtyards were used to light basement spaces that normally have no light or view. The employee's cafeteria has its own beautiful courtyard.

The structural frame is simple and repetitive. Precast concrete panels are used as a permeable skin with windows on both sides and a solid end condition where there are no windows. When we wanted a larger space to make the entrance, we maintained the whole structural frame, including all of the beams, so that the integrity of the structure to resist earthquakes is maintained throughout. I should add that this large setback, full of structure, is quite beautiful.

The exterior enclosure system in this building illustrates some theoretical isues about the nature of a curtain wall that I think are important. The quality of a thin envelope is not in the materials but in the design. In this case, it is made of precast concrete with glass infill, so it is actually quite heavy. Visually, however, it is a thin skin. The absence of modulations gives it the quality of surface tension characteristic of a skin. This is emphasized at the ends of the main facades by allowing for the actual 5-inch thickness of the panels to be visually evident. We were very interested in exploring the expressive possibilities of concrete in this building. Ever since Le Corbusier started using "beton brut," concrete has been seen as an artificial stone, made with stones, and has been employed by architects in one of two ways: either with a very rough surface texture, to show its aggregate, as in Paul Rudolph's work, or with an extremely smooth finish, to look like monolithic stone, as in I. M. Pei's work, and that of

Louis Kahn's. But concrete has another quality that may be more important, which is that it is a plastic material, a fluid that is poured in a container to harden. That is what we saw as concrete in this building; the forms are expressive of this plasticity and the quality is emphasized by the glossy epoxy paint finish (Figure 6–23).

The Museum of Modern Art project in New York, which we are working on now (Figure 6–24), is a major expansion and renovation of the existing museum plus an apartment tower. The project is very complex, involving the balancing of many competing factors. We have had to deal with preserving some buildings, adding new ones, combining functions and making it economical to operate and build. The wishes of a great many people and institutions had to be accomodated. In reorganizing the entry level, we have considerably increased the area of the zone available to visitors before they buy their tickets. The second floor, which is the same as the third, will have greatly increased galleries and a new system of escalators.

We made many studies of how to reconcile the design of the new street elevation with existing buildings alongside. The facade is very lightweight, 70 percent of its area consisting of opaque glass spandrel panels in a number of colors. This provided us with a framework to explore a variety of ideas about the nature of the wall in relation to the organization of internal space. We wanted apartments with small windows—a rather traditional idea—so that these windows would become special episodes in each room. Most of the museum walls are solid, without windows. The composition

6–23

6–24

of the building is in some ways reminiscent of a more traditional type consisting of a distinct top, middle and bottom. The base is the Museum, large in scale; and the middle is itself divided between lower and upper parts, within which one can sense floors, apartments and even component elements of the apartments in each floor; and the top is terraced. This hierarchy of parts is expressed as pattern on the facade and is graphic rather than modelled. An earlier architecture of yurtas and huts used similar devices. Several colors are used to create a rich composition within a very tight plane. This approach allows a minimal surface area to volume ratio, while at the same time accomodating a wide variety of scale-breaking sizes and parts. The composition of the facade will be enriched when reflections appear on the finished building.

The Niagara Falls Wintergarden (Figure 6–25) was designed and built for the city of Niagara Falls, New York. It is a very large, public greenhouse in the center of the city. The hope is that this project and others the city has been investing in will bring people back to the city. I believe that more than a building like this may be necessary. But meanwhile, we did the best greenhouse we could. The intention is that hotels and shops are going to attach to this pre-built atrium.

The Wintergarden is a marvelous place when Niagara is covered with snow. The trees inside remain green and the water keeps on flowing all through the winter.

I have talked a good deal about the concept of a skin, and some of the examples I have used,

6–25

6–26

like the Niagara Falls Wintergarden, are transparent. There are historical precedents for such an approach, particularly in vernacular architecture, where the skin is opaque. Good examples of this are the stave churches in Norway. The type (Figure 6–26) is interesting to me because it is all skin; forms are not an expression of a masonry tradition, but of sheltering an interior space. The forms are simple and yet extremely rich. The variation in the form of each church is considerable, yet all have a very simple structural system. As in the Wintergarden, they have a heavy interior frame with a skin separate from it, and supported by light secondary members, so that the church takes whatever shape is necessary for aesthetical or practical reasons.

Such correspondences with my own work were only discovered after my projects were completed. Nevertheless, I think the observations are valid, since they seem to confirm an underlying logic in what we have been doing, and establish for our necessarily new buildings a connection with the past that I regard as important for all good architecture.

6–27

7. Colloquium on Energy and Technology in Architecture

James M. Fitch: I found this symposium to be an extremely interesting event in which a number of authentically novel and valuable concepts were raised by the panelists. I'm certainly richer and wiser for having heard these arguments. Some of them, superficially, seem quite contradictory. For example, if you contrast Ralph Knowles' approach to the establishment of building to that of Cesar Pelli, you might think at first glance that there was some kind of mortal conflict between these two approaches. Actually, I don't think that is the case. I see no reason why, if he were so minded, Cesar couldn't model his forms according to the norms that Ralph has established. I don't think, in that sense, there is any fundamental contradiction.

Architecture is obviously an incredibly complicated problem, and architecture today is in a state of crisis around the world. It's not surprising that if you get six equally competent and intelligent observers of the essential problem—"What is wrong with architecture?"—they will view it from different vantage points. But we are all looking at the same problem, and bringing to this our own special expertise. My own feeling, of course, is that this has been a very productive program precisely because it has been organized around the concept of energy, and everybody here has brought some new perceptions to this problem.

I think, however, that we ought to establish certain benchmarks against which we can measure this stupefying scale of judgment, *vis à vis* energy. I think there are a few criteria in architectural design which are absolutely true. They are not relative; they are not matters of value judgment; they are absolutely true. For example, it is not a matter of opinion that water boils at 212°F at sea level. That is a description of experiential reality. It's not accidental that, in structural engineering, that member is described as best which does the most work with the least material. That is not a value judgment; that is absolutely true. And that is the benchmark against which we can assess or look at structural methods. There may be many reasons why, in a given project, your criterion is not to design the most efficient structure. But if you want to have an ultimate base against which you can grade these deviations, that's one of them. I think the same kind of criteria can be applied to any kind of mechanical system. Engineers of long ago established ratios of input to output, and these are not matters of opinion. A heat pump is absolutely a more effective way of concentrating heat in a certain area than is any other form.

If we remember that we do have firm bases on which to make judgments, then life will be a lot simpler for us as designers, and our decisions might prove to be more durable if we refer to these bases. So let's not consider energy a moral question. Let's say that certain problems, such as conservation of energy, can be described objectively. And if there is any future for our profession, they must be described.

I can't conceal my personal bias, and have to think that Ralph Knowles' work in studying the responses of buildings to solar energy is most enormously significant. We must realize that, whether or not we consider it important, the sun does, in fact, have this impact on the built world.

Sim Van der Ryn's work in the biological aspects of environment I find tremendously exciting. It is hard to believe that any young architect would not see that the parameters that are being suggested are liberating, not restrictive. They are not efforts to box you into mechanistic solutions. On the contrary, they show that a wide range of solutions is possible if you have the wisdom to recognize experiential reality, and don't allow yourself to be led into the enchanting *cul-de-sacs* which are offered by some.

Sim Van der Ryn: My message is simply that architecture and the liberation of architecture means that we have to begin to be involved with, and pay attention to, the whole constellation of support systems that make our quality of life possible. That is all part of the design problem. When I was first contacted about traveling to Ithaca, it was the people who were going to be here as part of this conference that interested me. To me, learning is what a large part of living is all about. I tend to organize my work not only around making a living, but also around learning.

When I graduated from high school, I received the prize in architecture, which was a copy of *American Building* by James M. Fitch. That book had an important influence on me, because even though I was European, the book showed that the roots for an American architecture were in America, and they grew out of the forces that shape this country—the climate and the people. And I've learned a lot from Ralph Erskine. His talk was a beautiful example of what architecture can be all about in human terms. His whole explanation of Scandinavian culture relates to the space and rhythm of weather, which is important. This cultural tradition reflects on our own experience, because we have never made those connections and do not have the kind of tradition that exists in Scandinavia.

Richard Stein's work, to me, is exemplary of the sort of creative work that architects can do. It's not design, in the sense that we are sold the psyche of design on a pedestal. Richard has done the important work in this country on connecting what architecture is as a physical and cultural artifact to energy consumption. More than all the rhetoric of the rest of the profession, Richard's writing makes real the responsibility that architects have in determining our destiny as a country, because what we decide and what we do sets in concrete the whole flow of energy in this culture.

There are many models for architects that are available, and, as Jim said, there's a liberating aspect to opening up our horizons and looking at all the various things that being an architect today can mean. We're not limited to being "gentlemen architects" designing irrelevant buildings for a small segment of society. There are all kinds of new roles that are opening up every day. Ralph Knowles showed us one. What is affecting about Ralph's work is his attention to a specific issue. The mastery and understanding and explication of the particular in architecture, and a particular which in this case makes a real difference (i. e., the sun), begins to give a level of understanding which goes way beyond a narrow technical knowledge of solar energy. It begins to give you a whole understanding of what forms are possible.

When I interview graduate students, I am only interested in one thing. What is it that they know really well? What can they tell me about that they really understand and care about deeply? Ralph, as an architect and a teacher, really exemplifies that. Architecture, to me, comes from a concept, but it also grows out of a mastery of a single element. My feeling is that if you can master a single element, and relate it to the whole, then you're at the beginning of your life as a creative architect.

Cesar Pelli's eloquent message is really very simple, at one level. The critical thing is to know what you are doing, and if you know what you are doing as an architect, you will always work with the vernacular. All the buildings he showed us are market buildings built in market conditions. They are not weird buildings for some weird client with an unlimited budget. They are the commercial buildings all over the country for clients who are interested primarily not in aesthetics but in having a workable product at a price they can afford. Cesar shows us that we can produce that kind of building by mastering not the esoteric, but the vernacular, and putting that together with a concept and an idea.

Ralph Erskine: There were three things that interested me here. One was just the very subject matter, which I thought excellent: summarizing some of the very important factors that I feel we, as architects, ought to be deeply concerned with. The second was to meet the people here—people I have read about. And the third was to visit this country, which I haven't done for a long time. I walked about New York a bit, and I'll take the opportunity to look at this part of the country before returning to New York. It's far too short a visit and no deeper understanding arises from that, but it helps with discussions and arguments and perhaps with preventing some misunderstandings and gathering new insights. I feel that this symposium certainly for me, has been very worthwhile.

Richard Stein: The necessary and almost obvious thing to say is that it has been a very informative and extremely worthwhile experience, having been here and participating in the whole discussion. Everything that has been said specifically recapitulating the intention and the direction of the different talks I can only second.

But beyond anything that it has meant to me, I think there is an enormous consequence, value and importance to the symposium. For a long time—for several years, certainly—the architectural profession has been floundering and searching for things that have been troubling it quite deeply for at least a decade, and the podium has been pre-empted by a group that has put together a very beguiling but eventually misleading set of formulations. I'm talking not only about the New York Five, but also about many others who have been proposing that the only significance of architecture now is to play a formalistic game that may be historic or allusive or vernacular, in which the form becomes the subject for discussion, and that the form can be divorced from the entire important responsibility of building. The unity, amid all the diversity, of the talks here has been the re-establishment of the fact that architecture primarily is concerned with building environments in which all our human activities take place. That is not something that is personally conceived by any of us; it is the total experience of the people who use it, the people who create it, the people who construct it. It represents our cultural values, and if the architecture profession is going to have the historic importance that it has had in the past, and should have in the future, it's going to be because the issues that were discussed here have become the subject of discussion. I don't think it's necessary that we all describe it in identical terms, or that the particular direction from which we move into it happens to be exactly the same avenue. It's almost like Camillo Cite's studies of the design of the square in front of the medieval churches: you can approach them many ways from many experiences, but all of them lead you into an understanding of the major edifices that you are approaching.

In the same way, the underlying current is that the problems of architecture are problems that have to do with the reality of how the building represents the most effective use of our diminishing resources, especially energy. All of these problems involve the things that shape the building, and these are the things that are the basis for discussion. Whether the solution happens to be a reflective facade on a building or the articulation disclosed by some important factor, they are factors that are concerned with building, and not with metaphors for building. We are not building a discussion, we are building a building! That was the underlying theme in all of the talks. If the discussion can continue to approach and resolve these problems, the intention of this symposium will be fulfilled.

Ralph Knowles: I agree with what Dick is saying about building. There was a book written some years ago called *Architecture of Space,* and there have been books written about architecture as form. In fact, there have been books about architecture as just about everything about the materials of which architecture is made, which, after all, are the materials of building. This is the fundamental issue that has been focused on here.

Anybody in the studio of Cornell or any other school has, at some time or another, been confronted by the question, "What is one handling here, when one designs a building?" It's in that regard that I found a lot of interest in what Cesar Pelli was showing: a different kind of energy conservation. For about three or four decades, we've been operating on the basis that there is value in efficiency. We tried to get more for less by adopting an attitude about technology that said, essentially, that if one could somehow put the machine to use it could reproduce like parts, and out of that we would get efficient architecture. Presumably, efficient architecture would conserve energy. We did that by looking at parts of the building in quite specific ways and we gave them specific functions, so that the notion of efficiency was based on a nineteenth century idea of technology which saw efficiency in terms of a single functioning of single pieces.

But what has been focused on here is a different sort of physical issue, representing a different kind of philosophy. It's the multiple use of a given piece of material. How many ways can you get it to function simultaneously? In that regard, it is interesting to look at the way nature functions and wonder whether, in fact, nature functions very efficiently. If you measure the efficiency of the functioning of nature in industrial terms, she looks bloody inefficient! For example, one of every hundred green turtles that are born survives. That sounds terribly inefficient to me. On the other hand, quite a lot is established from the 99 turtles that don't make it.

Looking at the slides of Cesar Pelli's buildings, I was thinking about how much was gotten from any given piece of material. An image, a reflection, enrichment, thermal property, transparency, translucency—all were gotten. Any given piece of material wasn't viewed in terms of any single function, but in terms of a multiplicity of roles.

I saw Erskine's work in exactly the same way. A given piece of material functioned in a variety of ways, and the notion of specialization of part in order to get the maximum efficiency. I think I saw that sort of thread running through everybody's work.

When you look at what you are doing in the studio, a more relevant question, it seems to me that what is certainly consistent with what has been said here is the manipulation of the material of construction, the material of building. How much are you getting for it when you put it in place? It's a question that interests me a lot, and I keep being reminded of how much good designers can really get for that material they are putting in place, and whether you think about that when you put the material in position, at least on paper.

Cesar Pelli: I feel very fortunate to have been at this conference. I enjoy discussing issues of architecture and design, but I enjoy much more the kind of discussion we are having now. Some students asked me if it was strange for me to be included in this group. It was not strange at all. It is very important that we get together, because a range of attitudes and approaches to problems includes the question, "How do we cope responsibly with the actions we have to take to make a building better?" The response requires us to be as aware as we can of the external circumstances and the way we can affect them. Architecture, sooner or later, has to catch up to reality, so the closer we are to it the more important and the more relevant our architecture will be.

Martin J. Harms: I think it's true to say that one of the particular interests of the people at Cornell would be design, but also what one might characterize as the formal issues of design, which aren't necessarily the same as "formalism." You see, an argument could be made that what Ralph Knowles is talking about is a form of environmental determinism. One could say that what Ralph Erskine is talking about is a kind of socio-political determinism. Does either position really take account of the difficulties associated with formal invention in architecture? The issue that a student, perhaps, would want to have answered is, "Where do architectural ideas come from? How do I solve problems of design? Where do imagination and metaphor and all those other kinds of issues, which in the design process are really independent of immediate environmental or social relevance, come in?"

Cesar Pelli: I think a forced conflict has been made between the elements that determine architecture. Actually, they don't determine, they set parameters, and I think that's an enormous difference. I think what Ralph was talking about was not determinism, but about certain parameters that one should take into account, although those are not the only ones. I believe it is a responsibility of the architect to work within those parameters, and to understand them. And even if you follow all of them, you're not going to get even halfway to architecture. That's why theories about design, attitudes about form, metaphor or whatever source of control you may have for making decisions about aesthetic issues that go beyond responding to the problem are needed. Cost, form and compliance with functional or energy criteria are not separate decisions, but really one decision that you are making at the same time. If one is good at that, one makes all of them good decisions.

Richard Stein: I think that the basis for the aesthetic that we apply to our architectural judgments is not something that can be completely dismissed. It comes out of certain attitudes, priorities and points of view, and it isn't timeless. The aesthetics of a period keep shifting. However, there are certain realities about architecture and the way it's built that makes certain very specific demands on form. They are not things that are said arbitrarily. There are certain ways in which space is spanned, space is enclosed; there are certain performance requirements that we expect of our buildings. All of these make very specific demands on appearance. These are things that aren't up for grabs, or things that can be dismissed or accepted. They are there, and they have to be understood.

Part of what becomes the aesthetic of building comes out of the knowledge, the very deep knowledge, of all of these elements, where the manipulation of the choices that are made within the acceptable range of parameters for them are so well understood that they can be put into the most favorable relationship: where they work most effectively, where they do the most for least, where they begin to establish the kind of tensions and tautness in building that, for me, are among the characteristics of aesthetics. I find that the question of formalism, of how form is derived, is essential. I don't think it's unimportant; I don't think that a building that just solves programmatic requirements is necessarily going to be a good building. Now we have such unlimited choices of material and ways of doing things that it becomes more important than it was (say, in the medieval period) to understand this and to make choices and intelligent decisions on where we are going. The choices available to a fairly primitive society in Balo Balo or someplace like that were so restricted that just understanding how to build a shelter created its own form. Now, when we have unlimited choices, we have to have an attitude about how to use them. I agree with Cesar that the importance of having a theory that is tied into the reality of what we are doing becomes essential, unless we are just going to flounder and hope for accidentals. I don't think that produces architecture. What we are looking for ought to direct a discussion toward a common acceptance of a general direction that includes a great plurality of choices, but not one that's tolerant of an infinite series of personal statements just because we can make them. The total sum consequence of these decisions has too much importance for the visual appearance of our built world and even for the economic well-being of people. The formal issues are very important.

James M. Fitch: In architecture, we make a very consistent error, based, I suppose, on our reliance on the printed page. Assuming that the aesthetic response is exclusively a visual phenomenon, if you look at the literature of architecture and take, for example, Mies Van der Rohe's famous pavilion at the Barcelona Fair in 1929, that is a building which had an authentically enormous impact on the whole development of the architecture of the period. It was really a shot heard round the world. And yet, if you examine the circumstances, you begin to realize that hardly anyone ever saw or experienced that building. I've only met two people who were even in Barcelona that summer!

Actually, its great reputation is based on a single set of photographs, and we don't even know any more who took the photographs. One set of them, after countless reproductions, now remains in the Museum of Modern Art. The reputation of this building is based exclusively on black and white, two-dimensional facsimilies for four-dimensional reality! Chances are that the building in experiential reality was not only less effective than we have assumed it was on the basis of these photographs, but that it was almost certainly disastrously unsuccessful from an environmental point of view. If you look at the photographs, the first thing you notice is that there is not a living animal in the whole field of vision. There is not a gull, a cat or a man; there's no life in those pictures. Part of this is because architectural photographers don't like people, since they disturb the composition. But if you know anything about Barcelona summers—and these photos were taken in the summer, at noon—you know that this hermetically-sealed pavilion in marble and glass was probably intolerably hot. The reason that there is nobody in those pictures is because life couldn't be sustained in that kind of environment.

It's impossible to imagine architects today getting along without photography, but it's also quite possible to argue that photography is the worst thing that ever happened to architecture. It's very important that you try to make corrections for these facts.

Another example of this is Falling Water, the great house that Frank Lloyd Wright designed for the Kaufmann family. Every student of architecture could draw the house blindfolded. But we don't ask ourselves *why* this thing was imprinted on our retinas, what special aspect of this building was imprinted, and from what special vantage point in time and space it was imprinted. Classic photographs of the house are taken from down below the house in the valley, around noon. The absence of trees indicates wintertime; the fact that the water is still falling and not frozen indicates that the temperature is not below zero. Actually, photography gives you a tunnel view of the reality of that building. If any of you have been fortunate enough to go there in the summertime, you realize that the valley is full of poison ivy and copperheads; nobody ever went down in that valley except the photographer. Kaufmann certainly never sat down there and enjoyed the building from that point of view. On the contrary, the family was in the house. And if you go inside, you find that the rooms have low ceilings, they are quite dark, the humidity is very high and, most of all, the roar of that damn waterfall is stupefying!

If the next generation of architects wants to be more successful than we have been, they will have to deal with corrective factors, and the total reality as it is actually experienced.

The printed page and photographs produce misconceptions so monumental and so fundamental that it takes a lot of effort to fight your way out of them. You can't get along without photographs—but don't trust any of them.

Ralph Knowles: There is another side to that coin. I heard Bob Stern speaking at Harvard in 1978 at a conference that was concerned with introducing energy-related issues into architectural criteria. He made the point that, among other things, there was a kind of immorality practiced for a period of time that essentially threw much of history away. The Bauhaus and notions about industrialization occurred during a period of incredibly rapid growth and uncertainty, during which we were inclined to throw away the deep roots of much of our architecture and replace them with a different attitude. The Barcelona building was the beginning of that. But Stern pointed out, correctly I think, that concerns with environment, context and energy were not new. He pointed to architects like Richardson and others. A short time later I happened to be in Chicago, where I went to see a building by Richardson that I had seen photographed many times, a house in the familiar Richardson style, with a stone, almost impervious, facade. I thought that was what Richardson was all about. I walked through the front door, which was a tiny hole in the wall, into a lobby which rose a certain distance, and beyond that was an almost totally glass wall facing out onto a garden. I hadn't had the slightest idea that *that* was what he was about. I then went to the Robey house, of which I had also seen many pictures. A photograph of that house must have influenced more young designers than any other single photograph except those of the Barcelona Pavilion and Falling Water. We all had to make those overhangs and reproduce those balconies and the flowerpark at the end of the building. The walls had to overlap and turn a certain way. We learned from that house, and that photograph of it. The other bits of information from which we learned were the plan drawings of the building. And from that photograph and those plan drawings, I thought I knew everything about the Robey house. The fact is, I didn't. I walked up the stairs into the very long space that makes up what you see under the roof and it turned out that while it looked symmetrical in the plan, and one assumed it was symmetrical from the photograph, it wasn't the least bit symmetrical! The windows on the north are only half the size of the windows on the south, and the overhangs are different. Except for that visit, the direct involvement, I would forever have had a misconception of Frank Lloyd Wright's houses.

The parameters that Cesar talks about are, in fact, the real stuff. You have to really experience them. One of the incredible weaknesses of studio education is that it's in the studio, not in the world. The more contact an architect can have with his surroundings, the better. They are the real point of departure for a student. Beyond that, he needs to have more contact with actual buildings. A photograph, the two-dimensional image of a building, conveys a certain amount of information. Unfortunately, it's a very efficient way to transfer information of a certain sort, but not very useful architecturally.

James Fitch: Information that a photograph gives you is not necessarily incorrect. It's just grotesquely limited. It gives a tunnel, pinpoint view of experiential reality. If a photographer takes a photograph of a house and he's standing on a flat plane, there are 360° horizontal and 180° vertical, with an infinite number of points between each degree. The photographer has to choose one of those points. If you take the question of time and multiply it by the minutes, hours and days of the year, and then multiply it by the number of points in space the photographer has to choose from, you come up with something like 2,485,367,000 points in time and space, one of which—and only one—is selected by the photographer. With this limitation, quite obviously, the information conveyed is also limited.

Question From The Audience: I would like to know what those issues were about which you gentlemen couldn't agree.

James Fitch: One of them has to do with Cesar's building in Los Angeles, but it also refers to thousands of buildings in the world today, whose walls are thin membranes, very transparent, with almost instantaneous heat transfer. The point is: these buildings stand in the real world, which is asymetric. The path of the sun, as you saw from Ralph Knowles' diagrams, is anything but symmetrical in time and space. A southwall in Chicago during the year will receive 79 times as much insolation as will a north wall in the same building. In mid-summer, the south wall of a building in Chicago will receive 189 times as much heat as will the north wall. That suggests that any free-standing tower like the Hancock Tower in Chicago does not stand in a thermally symmetrical environment. The discontinuities across the 120-foot deep dimension of that tower make a temperature differential that you would have to go from Chicago to Miami Beach to make in the outside world. Actually, I think you would have to go beyond Miami Beach. This means that, inside that 120-foot depth of the tower, you are trying to correct by purely mechanical means a thermal dimension of 1500 miles! As far as the thermal disequilibrium is concerned, that building is 1500 miles deep, not 120 feet.

Yet, our ambition is to make every square foot

of each floor equally comfortable. With modern technology we rely exclusively on mechanical means for achieving that equilibrium. My question about all thin-skinned buildings is whether or not this is an optimal solution: whether or not the architecture itself shouldn't take some of the load of correcting this disequilibrium before we call for the sophisticated systems. Another building that Cesar Pelli showed us recognized this fact: the north wall was dramatically different from the south wall visually, because it was different thermally. These are two good examples of how we ought to approach this question. I feel that this is one of the great errors of contemporary architecture. Because we are possessed with formal ambitions for aesthetic success, we really produce uninhabitable spacc and then turn it over to the engineers to make it habitable. An historic function of the architect is to produce habitable space. Until 100 years ago, there wasn't any G. E. compressor onto whose shoulders we could shove this responsibility, so we were compelled to solve this in architectonic terms. I think it's a great mistake to use modern technology to escape this obligation.

Richard Stein: One of the subjects on which there is some disagreement is how one makes certain assessments. What are the priorities of the total maker, upon whom these issues are dumped, and from whom we get the solutions? I'm interested in the fact that there is a basis for discussing these things. That is a tremendous triumph. But there is, for me at least, often a distortion of the question being asked. Possibly this is because I don't see the problem in the same terms that someone else would. Specifically, I was having some difficulty understanding the application of the solar envelopes that were so beautifully developed and prepared by Ralph Knowles. It was a superb analysis of what happens if one makes that the single parameter for determining decisions. However, the solar impact on a building affects it in many different ways, and the ensuring of solar access to the neighbor's building often is a penalty carried by the building within the solar envelope itself. That seems to make the resulting form a form that is derived by what is, in many cases, a secondary or even tertiary problem. If one understands buildings as complete mechanisms that have many sorts of requirements, one of which is solar access, in many buildings it may be more important to reject rather than to provide solar access. Under certain conditions, solar heat may even be damaging. Others are problems of lighting, space and ambience, psychological connections to the outdoors and problems of depth of building.

What's interesting is having a prime demand emerge out of the kind of energy-related requirements we are talking about. The narrow, wing-shape of Pelli's U. S. Embassy building in Japan evolved out of a complicated and difficult set of requirements. I think it was handled in an interesting way, but I'm not sure the game that was played between the facade as a skin and the inside was the only way to handle it. What came up as a basis for discussion was this: How does one evaluate a set priority under the number of conditions that apply to the design of the building? It's important that we do not have such a myopic view of the problem, or allow ourselves to be pushed into such a specialized order that we fail to see the enormous complexity of the decisions. Out of the humble response to this enormous complexity, buildings truly find their form and shape. These are the ones that become, in time, the vernacular buildings to which we look back.

Ralph Knowles: One of the areas that Dick and I were discussing was whether or not the notion of the basic right of solar access flies in the face of a general sort of cultural condition, and even in the face of urbanity itself. There are a couple of ways of looking at that. First, we viewed the envelope as a possible expression of the public will through public policy. That is, it would not happen unless the public itself values access to the sun. That is true of much of what we have been discussing. One of the contexts within which the architect works is a context of public value, and that is a context with which you are pretty much stuck. If the public does not value access to the sun, the public is not going to get access to the sun. In my part of the country, it is a condition that is increasingly valued, for one reason or another. It may, in fact, be quite different in New York. The whole sense of the city, the whole sense of what is urban, is different there. The question that Dick was raising is whether one could not modify the notion of the solar envelope in order to come closer to fitting a more universal image of urbanity. At that point, Cesar suggested the notion of the solar envelope as I presented it, as an extraordinary undeveloped notion. It is just the beginning, based on a sense of public right. In the expression of the idea, it's perhaps oversimplified at present. But in its actual application, it's going to have to respond to complexity and take a lot of different forms. It will be modified again and again until it, as a concept of public right, comes into balance with the public concept of what is "city." I don't have the slightest idea of what that ultimate modification may be. I think that is to the good. If I could conceive of a single set, a single framework, and impose it with a free will on the world, I think I would quit.

Sim Van Der Ryn: I don't think we ought to apologize. Let's face it. The high-rise building is the cancer of modern architecture. I think it works very well in some places, and I think we ought to maintain places like Manhattan, and its monuments from that era. But, basically, it's insane. In terms of all the conditions we talked about, high-rise buildings are simply a monument to man's greed for land; they are simply a macho match between male architects. I think the high-rise building is an important part of that adolescent pitching match that the twentieth century has been through, and Cesar gave some of the reasons that made it possible. But in this symposium we have given a lot of reasons why it is not really a good way to go any more, in intimate terms, energy terms and ecological terms. I agree with Ralph that the solar zoning is the beginning of the tool, and I would not want to see it get interpreted literally. In California this year, one of our legislators who is very enthusiastic about solar energy, passed a bill which said that any tree or piece of vegetation that interfered with a solar collector should be cut down. That's an example of the dumb legislation for solar zoning that we have to guard against. But it represents why looking at the environment in this dynamic way is really so important. I know in my own work in Sacramento, where I kind of played monopoly with 40 square blocks, that one square building 15 floors high totally constrains what we could do on 6 or 7 surrounding blocks. The whole value of the sort of analysis Ralph is proposing made absolute sense. But I really don't like high-rise buildings. I don't think they make sense in the kind of culture we're talking about.

Richard Stein: I guess somebody has to come to the defense of high-rise buildings, and I propose to do it. Not only do I live in New York, but I live there by choice. I live in New York because the high-rise buildings and the number of people that they bring together gives an enormous variety of choices and an enormous variety of opportunities for what one can do. My office is within five blocks of the museum that Cesar is going to extend into a national monument; within five blocks of a public library; and surrounded by six book shops. I can find more wonderful places to get Japanese sushi than Tokyo has. The things that happen as a result of having enough people so that all the exotic and special tastes of any of them can find expression is part of the whole interest in urbanism. The growth of metropolitan regions—even though people say, "Oh, I live in the suburbs," or whatever—is because the concentration does give this kind of cultural quality an opportunity to exist.

Also, surprisingly enough, within our present mode of living, it turns out to be a relatively efficient energy expenditure by comparison with other choices. It means that instead of people having to drive their 12-mile-per-gallon car to meet somebody in a restaurant for lunch, they just take the elevator down to the street. All of these things give you an idea of the complexity of the design decisions that we are dealing with, and the difficulty of making these snap judgments. I don't think that the whole world ought to be created in Manhattan's image, but I also don't think it ought to be created in the image of, say, two-story suburbia.

I think one of the difficulties of saying that the most important characteristic is the protection of one's rights to sun is what it predicates. If each person is allowed to build within a prescribed urban density, which is then considered an unalterable right, it states that this is the only density that will be allowed there from now on. It prevents the kind of accumulation that is sometimes important and required by activities. Suddenly we find the law condoning what, in certain circumstances, is actually an anti-social pattern of behavior. That's the reason I was very interested to see the implications of the solar envelope studies. It doesn't mean that the importance of the sun is to be made any less of.

I made a remark earlier and would like to mention it again, because I think it puts the idea of solar protection in a different light. The whole environment we're in, even when it's 0° or 30° or 50° outside, is that much above absolute zero because of the sun. The sun's impact is not only from the direct rays we receive or don't receive—it's the whole creation of a generally satisfactory environment that has a differential we can deal with. Other aspects of the sun, such as the air movement that comes from the sun's response to different reflective conditions, are important, too, and in turn are probably at least as deserving of consideration as the shape of the enclosed trapezoid that would be created by the path of the sun's movements.

The problem is one of putting in perspective all of the factors that impinge on the creation of environments for activity.

James M. Fitch: I'm sure Ralph Knowles never visualized that if his solar access zoning was adopted it would be retroactive; that is to say, that any town that adopted it would be razed to the ground and built over again. I'm sure no zoning ordinances ever envisioned that. And I'm astonished to hear you say that you think this kind of zoning would be anti-social. It seems to me the motivation is highly social; namely, that all of us have equal rights of access to the sun. It may very well be that this would work out to the disadvantage of an individual property owner, but this is no moral argument against the overall proposition. Almost any legislation aimed at raising the general level of well-being is apt to cause hardships for the individual whose private plans don't agree. Any type of zoning has that hazard, so I think it's really a rather limited attitude to say that the consequences of this could only be negative.

Sim Van der Ryn: I'm surprised to hear Dick Stein say that he connects high-rise buildings with either sociability or urbanity. I think what makes New York urban to me is that people are quite common there, as opposed to the car. In looking at the question of density, as soon as you begin to accept the automobile as pollution in an urban area and begin to design it out of the urban core, you find that you can create very sociable, urban, high-density places, without resorting to high-rise buildings. In Sacramento, when I proposed the idea of no new high-rise buildings, what I really wanted to do was test that notion of density. My feeling is that high-rise buildings are anti-social, and not very comfortable either.

Richard Stein: Then why do you live near San Francisco, rather than Sacramento?

Sim Van der Ryn: I go to San Francisco very rarely. I don't live *near* any place, I live where I live. But I sure wouldn't live anywhere because it has high-rise buildings.

In the 20 years that I have lived in the Bay area, I have seen San Francisco transformed from a very urban place, where there were few buildings over 6 stories, to a place where the dirtiest word you can say is Manhattanization. Like the Oregonians who have bumper stickers that say, "Don't Californiaze Oregon," we say, "Don't Manhattanize San Francisco." And if you want to talk about solar rights, the entire waterfront area of San Francisco has been blocked by high-rise buildings which not only have changed our solar pattern but also our view pattern—in fact, have changed the whole form of the city.

What happened in San Francisco, and really contributed to this unique urban form, was the fact that the typical pattern was comprised of light masonry buildings no higher than four to six stories. Only on the hill did you have higher buildings, conforming very much to what Ralph was talking about. Now, in 15 years, the entire face of the city has been changed, and the character of the city has changed. I was born in New York and grew up in Manhattan, and I certainly connect Manhattan with this particular pattern of high-rise buildings. But I don't connect urbanity with that particular pattern.

So, when I attack that kind of urban form, I'm not attacking urbanity. I'm saying there are whole realms of options, all kinds of ways of creating a truly urban place, a high-density place, without having to resort to high-rise buildings, which came to us because they were technically possible.

Question from the Audience: Can you name some high-density urban places that do not go over four to six stories?

Sim Van der Ryn: Yes. Amsterdam, Rome, Zurich, Los Angeles, Paris and most of London, to start with.

James M. Fitch: Those cities were urban centuries before high-rise buildings were built.

Sim Van der Ryn: High-rises do have their advantages, when society is organized in a certain way. When people are living 40 miles out and commuting to the city and their specialized jobs on the fortieth floor, they have advantages for that particular organization. But I can't accept what they represent. I remember coming back to New York in 1957, when jobs were hard to get, and I was hired as a designer at Skidmore and Merrill. I traveled in from the suburbs, and went into one of those buildings at 8:30 in the morning. At 9:00 it was coffee time, and I walked out of the building and never went back in. That was not my idea of how I wanted to live, and it still isn't.

Richard Stein: That indicates the wonderful diversity of choices. I don't begrudge anybody the lifestyle that he or she selects, but I think that under some conditions and circumstances, we have an available set of technological tools that can be misused or properly used. In all the things we do, we have the choice of making either intelligent or stupid decisions. We don't base our whole philosophy on an analysis of the stupid decisions. We don't always talk about the idiocy of rural life; we also say that it has wonderful qualities. And even the greatest of the self-contained communities have an umbilical connection to the people who produced the Volkswagen that provided the fenders to build from, or the people who produced the waste products to build the throw-away building. The thing to do is to edit diversity, to use the things that can be used more intelligently than they have been used before.

Ralph Knowles: There are a couple of ways of looking at that. I think that at the present level of technology of active systems, they are not very useful. They are also, in a broader sense, not very architectural. They are, in the main, something added to buildings rather than something inherent in buildings. There is no tradition—there are no deep roots connected with that sort of system—as there is in relation to what we call passive systems. We've got centuries to look back on to get better information from, to learn from. That's a particularly valid point if you're looking at active systems on a tiny scale; that is, one house at a time. If you look at active systems on a very large scale, they are frightening. They are just as frightening to me as the large-scale energy supply systems we presently use. They represent a degree of centralized control that doesn't essentially make them any different from any other highly centralized system. And with that high degree of centralization, they are very susceptible to failure or breakdown, as are other highly centralized energy supply systems. Those two factors worry me a great deal. The ultimate sort of centralized active system is the pie in the sky that collects the solar energy and beams it down to earth. I have real problems dealing with that. I'd like to know who is going to own that pie; who will control it; and who is going to try to shoot it out of the sky? I don't like those kinds of problems. They are beyond my control; they are beyond my comprehension; and they endanger me. I want the control over my system that modifies my environment to be as close to my control as possible, as close to my policing power as possible. I see a very great danger in large-scale active systems, and for the moment I think small-scale active systems are interesting but totally inadequate, and, to a great extent, they are beside the point of architecture as a process.

Question from the Audience: Can these criteria for your work be applied to larger constructions than domestic dwellings?

Ralph Knowles: Nothing would prevent it. I am not opposed to making large things. In nature, there are certain kinds of limitations on size. A number of times people have mentioned quite an old book by D'Arcy Thompson called *On Growth and Form.* And there is a chapter in it called something like, "On Being the Right Size." How do you know that you are big enough? How does the whale know that it is big enough? There aren't many things larger than a whale, or a redwood. There aren't many things larger than the largest building, but I don't know how big the largest building can be. Like Sim, I would like to find out. I would be interested in pushing against the upper limits of size in order to find out about self-sufficiency. I think we know very little about that, partly because of our economy, partly because of our technology, partly because of our culture. But the fact is, we do know big things, and we have built bigger things. I'm interested in the biggest thing we could build, and pushing against that kind of limit to find out what it is about. Unless we push against it we're not going to know. Whether it has to have only negative qualities is a debatable point. I'm not sure that big is necessarily worse, and that small is necessarily better. I think lots of big things we build are bad, but so are lots of small things.

James M. Fitch: Ralph, I'm surprised that you had to refer to earlier work you did five or eight years ago, in which you were concerned with exploring the proper response of isolated big buildings to solar zoning. If I understand the logic of your work, you exploited that to a certain extent, and then began to realize that in real life these buildings don't exist in isolation, that the context also had to be subjected to the same scrutiny.

Ralph Knowles: One could take the attitude that in large size lies a certain freedom to shape a response to certain kinds of environmental factors. One could generate a large shape in relation to the dynamics of size, in order to control with great precision levels of irradiation by day and by season. One really can only do that with very large forms, because with small ones the increment with which we are used to building, the room, the space, gets in the way. It's like trying to sculpt a portrait with an increment of shaping that is the same volume as the portrait. If that increment happens to be cubical, then we are going to make a block head. As you reduce the size of that shaping increment, the generative increment in relation to what you are trying to make, you gain shaping freedom. As you reduce it again, you gain more shaping freedom, until finally, at some point, the increment of generation becomes small enough so that you can produce a refined portrait of the face. It would help if some increments were finer than others: I can sculpt the back of the head in a rough sort of way, but around the eyes and mouth I have to have a fairly refined increment.

Sim talks about San Francisco at an earlier time. What happened there is that the increment with which the city was generated was relatively small in relation to the general configuration of the city. The hills went up and down, and the incrementation was able to follow that. That was the sort of ideal I was working with a few years ago, except that what interested me was the possibility of shaping a huge building that gained flexibility of shape by the fact that it was very large in relation to the generative increment. I've stopped pursuing that, partly because of the reasons Jim suggested, and partly because for the moment that is not a real problem. If I'm going to contribute anything as a teacher and a researcher, I've got to confront problems as near as possible—not yesterday, and maybe not today, but not so many days in the future that they can't be of some practical use.

Sim Van der Ryn: Your talk about increments connects to my earlier diatribe, and should be connected to what Cesar was saying, which was extremely profound. One measure of a democracy, one measure of participation, is the number of people who can commission buildings. To me, there is a direct relationship between a San Francisco, which was created out of increments of four to six stories, which were controlled and built by local people who lived there, and a Manhattan, which has increments representing investments of hundreds of millions of dollars, which are controlled by corporations that aren't accountable because they aren't people. That's one of the great fictions of the law. To me—and I grew up there—the problems of New York are directly related to the fact that the increment of size and economic decision-making is totally transformed. What that represents to me is the paradigm of the high-rise building. As it moves across the world, it represents that whole change—whether it be in Lima, Havana, Caracas or San Francisco. The actual architecture is only a small part of it. I don't think they turn out to be very good environments, and they represent a particular way of looking at space and what people do in that space. But more than that, they represent a particular aspect of the economy and of control which is then reflected in a whole series of social consequences.

Richard Stein: I'm not interested in pursuing the greatness of the biggest building that can be built. That is not a problem that interests me as an architect. I am interested, however, in understanding the complete dymanics of buildings and how they operate. I would not recommend—in fact, I would take a strong stand against—the further densification of Manhattan. I think that becomes a counterproductive act, and that you begin to lose more amenities by building than you gain. Requirements to serve it, the new demands on infra-structures that have to be fitted into a crowded street, are so expensive, so unrewarding and so disruptive that at this point I'm not an advocate for infinite multiplication. On the other hand, because of the stupidity with which high-rise buildings have been built in the last 20 years, with the same building being built in Caracas and Montreal, I don't think the idea of urbanization, of people being close together, making certain connections and interactions, should be a completely negative one, despite the inanities of high-rise buildings. Some of the most enjoyable buildings in New York are some of the high-rise buildings: the Woolworth Building, the Empire State, the RCA . . . They are fascinating buildings, and the quality of space and view and involvement surely is as good as being on the third floor of a building in San Francisco.

Index

ARCHITECTS, ARTISTS, HISTORIANS, AND AUTHORS

BUILDINGS

GEOGRAPHIC AREAS